Agricultural Water Management

Premjit Sharma is having more than fifteen years of teaching and research experience mainly on cropping systems and soil fertility. Throughout his career, he has been associated with a number of educational and research institutes. He has published a number of research articles in reputed journals. His areas of interest include agriculture, agricultural economics, agricultural business management, natural resource management and sustainable development.

He has been associated with a dozen of UG and PG Courses in the discipline of agriculture and is the chairman of advisory committee of M.Sc. and Ph.D. students. Apart from being an active and dedicated teacher, he has also been committed in research and running projects funded by the eminent bodies.

Agricultural Water Management

Editor
Premjit Sharma

2026
Gene-Tech Books

ISBN 81-89729-23-3

Published by : **GENE-TECH BOOKS**
4762-63/23, Ansari Road, Darya Ganj,
NEW DELHI - 110 002
Phone: 43003222
e-mail: genetechbooks@yahoo.co.in

PRINTED IN INDIA

Preface

In recent years, agricultural water has helped to meet the rapidly rising demand for food, and has contributed to the growth of farm profitability and poverty reduction as well as to regional development and environmental protection. After several decades of publicly funded surface irrigation, and more recently of privately developed groundwater irrigation, remaining opportunities to harness new resources for agriculture are fewer and more expensive. Investment is increasingly focused on rehabilitating and improving the existing systems.

However, water productivity remains generally low, and returns to public investment generally disappointing, especially in large-scale irrigation. New solutions are needed, based on new management options and widely available technologies. The role of government is changing, responsibility is being decentralised, farmers are playing an increasingly important role in decisions and investment, and more and more, markets are driving growth. How to meet ever rising demand for food while at the same time increasing farmer incomes, reducing poverty, and protecting the environment, all from an increasingly constrained water resource base, is the main challenge facing agricultural water management.

Agricultural water management is not a goal in itself but part of a process of resource management that provides a key input to agricultural production and farmer incomes. It includes irrigation and drainage, water management in rainfed agriculture, recycled water reuse, water and land conservation, and watershed management. It covers all irrigated agriculture, whether fed by surface water or groundwater, including both public schemes and millions of private individually irrigated farms, in a wide range of agro-climatic conditions, and in a broad set of production systems and water management contexts. Agricultural water management is at the crossroads between four areas of public policy for sustainable growth: water resource management, agriculture, rural development, and the environment. It also interacts closely with broader aspects of macroeconomic policy for growth.

This book highlights the importance of agricultural water management and describes the advantages and disadvantages of conventional and innovative systems for the treatment and disposal of agricultural water. It would be a useful reference tool for policy makers, project managers, development organisations, researchers and other interest groups working for effective agricultural water management.

Editor

Contents

1

Potentials of Water in Agriculture

Almost all of the water of the planet occurs as saltwater in the oceans. Of the 3 percent of the global resource that is freshwater, two-thirds comes as snow and ice in polar and mountainous regions. Hence, liquid freshwater constitutes about 1 percent of the global water resource. At any one time, almost all of this occurs as groundwater, while less than 2 percent of it is to be found in rivers and lakes. In the temperate, humid climates, about 40 percent of the precipitation ends up in the groundwater, and for Mediterranean-type climates the figure is 10-20 percent. For the truly dry climates, the value can be virtually zero.

Not all the water in rivers and lakes and in the groundwater is accessible for use because part of the water flows in remote rivers and during seasonal floods that cannot be captured before the water reaches the ocean. An estimated 9 000—14 000 km^3 of water is economically available each year for human use. At most, this represents 0.001 percent of the estimated global water. At present, annual withdrawals of water for human use are about 3 600 km^3. This may give the impression that there is plenty of water available that could be withdrawn for human use.

However, part of the available surface water must remain in the rivers and streams to ensure effluent dilution and safeguard the integrity of the aquatic ecosystem. How large this part should be is little understood. It varies with the time of year, and each river basin has its own specific ecological limit below which the system can be expected to degrade.

A global estimate for this demand is 2 350 km^3/year. Adding this to the amount withdrawn each year for human use results in nearly 6 000 km^3 of economically accessible water that is already committed. This indicates that globally the margin is fairly small. Because water and population are distributed unevenly throughout the world, the water situation is already critical in various countries and regions and likely to become so in several more.

Agriculture is the principal user of all water resources taken together, i.e. rainfall (so-called green water) and water in rivers, lakes and aquifers (so-called blue water). It accounts for about 70 percent of all withdrawals worldwide, with domestic use amounting to about 10 percent and industry using some 21 percent.

There is an important distinction between water withdrawn for use and water actually consumed. In irrigated agriculture, about half of the water withdrawn is consumed in evaporation and transpiration from plants and moist soil surfaces. Some of the plants that contribute to this evapotranspiration process are unproductive weeds and plants on wasteland. Water that is abstracted but not consumed infiltrates the soil and is stored as groundwater or flows back through drains into rivers. This drainage water is generally of a lower quality than the water originally withdrawn owing to contamination by agrochemicals and salts leached from the soil profile.

Compared with a return flow of 50 percent of the water withdrawn for agriculture, 90 percent of the water for domestic use is returned to rivers and aquifers as wastewater, while industry typically returns up to 95 percent. Poor-quality return flows from urban and industrial areas are sometimes treated before being returned to watercourses, but the non-point character of agricultural pollution makes treatment difficult. In this sense, agricultural water pollution may be better handled by controlling quantitative use and outflow from agricultural land.

Significance of Rainfed Production

Rain is the source of water for crop production in the more humid regions of the world where some 60 percent of the world's food crops are grown. Rainfed agriculture takes place on some 80 percent of the arable land and irrigated agriculture produces 40 percent of the world's food crops on the remaining 20 percent. In order to meet future food demands, it is expected that relatively more crops will have to be grown on irrigated than on rainfed land, such that about equal amounts will come from both types of areas.

Given the importance of rainfed cereal production, insufficient attention has been paid to potential production growth in rainfed areas. Most attention usually focuses on the possible expansion of irrigated areas. However, increasing cereal yields in rainfed temperate countries, better plant protection and manure techniques, and the use of supplemental irrigation in more arid countries indicate the significant potential for improving rainfed agriculture.

In arid regions, water scarcity is the result of insufficient rainfall. Semi-arid regions may receive enough annual rainfall to support crops but the rainfall is distributed so unevenly in space and time that rainfed agriculture is barely possible. Rainfall variability generally increases with a decrease in annual amounts and it is particularly high in the Sahelian countries. These regions are known for their periodic droughts that may last several years. Rain in semi-arid regions also tends to fall in a few hard showers. Such rainfall is difficult to capture for agricultural use. This leads to large amounts of runoff going into drains and eventually seeping down to the groundwater or to rivers.

Where river discharge is large and difficult to manage, one way to capture the flow is through spate irrigation in which part or all of the river flow is diverted into fields surrounded by high bunds. In this way, one irrigation of up to 50 cm can provide enough soil moisture for a wheat crop even in the skeletal soils found in Yemen. Floodwater harvesting takes place within a streambed and entails blocking the water flow. This causes water to concentrate in the streambed. After the flood season is over, the streambed area where the water collects is cultivated. A terraced wadi (ephemeral stream) system is one type of floodwater harvesting. Here, a series of low check dams are constructed across a wadi and the wadi area is cultivated. Too much flow will breach the check dams or the diversion structures in spate irrigation. The suitability of these methods also depends on the soil conditions and depth of the wadi bed. Water harvesting, which is the collection and storage of surface runoff, has also proved useful in semi-arid regions with infrequent rains.

Although there is a great variety of such rainwater technologies, it is not clear whether their widespread use is always feasible, especially for poor farmers. The costs involved in the construction and maintenance of the water harvesting system play a major role in farmers' decisions on whether to adopt the technique or not. In the past, water harvesting systems were often installed with financial support from outside agencies, such as NGOs and international funding agencies. Many of those systems failed because of lack of involvement of the beneficiaries and their inability to organise and pay for maintenance.

Rosegrant et al. report construction costs for water harvesting systems in Turkana, Kenya, of US$625-1,015/ha. Labour and construction constitute the bulk of the water harvesting costs as the opportunity cost for using the land is essentially zero. The initial high labour costs of building the water harvesting system often provide a disincentive for adoption of the technique. Moreover, many farmers in arid or semi-arid areas do not have the human resources available to move the large amounts of earth necessary in the larger systems. Therefore, small-scale water and soil conservation techniques that are applicable at field level are often adopted more easily. Investments on a larger scale require the existence or creation of community organisations both to pay for the necessary investment and maintenance, and to manage the benefits of the water harvesting infrastructure.

Maintenance of the system is sometimes required in the rainy season when labour is relatively scarce and therefore expensive because of competition with conventional agriculture. Notwithstanding these reservations about the widespread applicability of extensive water harvesting systems, model studies indicate that there is significant scope for increasing rainfed production provided that appropriate investments and policy changes are made. Crop breeding specifically for rainfed environments is crucial to future cereal growth.

Significance of Groundwater

Groundwater use for irrigation presents a paradox: regions where the groundwater resource has been overdeveloped coexist with regions with considerable potential for development of groundwater for use in irrigated agriculture. A corollary is the so-called fallacy of aggregation: in aggregate terms, at the global or even national level, groundwater availability appears far in excess of present use. The annual groundwater use for the world as a whole has been placed at 750—800 km^3. This figure may appear modest compared to the overall groundwater reserves, but only a fraction of the world's groundwater reserves are economically available for agriculture. It is estimated that about 30 percent of the world's irrigation supply is made up by groundwater but this input accounts for some of the highest yields and highest value crops.

The number of tubewells providing water to irrigated land in India, China, Pakistan, Mexico and many other countries has grown rapidly in the past 40 years. For example, some 60 percent of irrigated cereal production in India depends on irrigation from groundwater wells. This has led to widespread and uncontrolled overabstraction of the resource and the creation of a 'bubble' groundwater economy. In Yemen, abstraction is estimated to exceed recharge by 400 percent.

Groundwater abstraction and recharge have rarely been quantified accurately. This should be a first step in assessing the potential for further development of the resource and designing management approaches. Where irrigated agriculture depends in part on pumped groundwater, many of the command areas present a mosaic of irrigation methods. These range from totally irrigated by canal water to entirely fed by pumped groundwater, with most of the fields having some combination of the two. Hence, irrigation is by definition conjunctive, but there are few examples of conjunctive management.

In China, 52 percent of the irrigated lands are (at least in part) served by tubewells. As a result of overabstraction of groundwater, water tables have fallen by up to 50 m over the last 30 years. For example, in the Fuyung Basin in north China, surface water has been curtailed drastically in order to meet industrial demand, and farmers have responded by resorting to groundwater irrigation. The root of the Asian groundwater crisis alluded to by Shah et al. that threatens millions of poor rural communities lies in the open access nature of the resource. Paradoxically, it is precisely this feature of groundwater in shallow aquifers that has made it a powerful tool in the fight against poverty, i.e. that everyone who can afford to install a pump has free access to water.

Irrigation with groundwater is generally more productive than canal irrigation because groundwater is produced close to where it is used with hardly any losses during transport. In addition, farmers are in control of the timing and amount of the water extracted. Evidence in India suggests that crop yield per cubic metre of water on groundwater-irrigated farms tends to be 1—3 times higher than on farms irrigated with surface water.

Throughout the world, most groundwater development has proceeded primarily on the basis of individual initiatives. Unlike surface irrigation or drinking-water supply projects where government agencies are generally involved in many aspects of design, financing and implementation, most groundwater development is driven by the decision of individual farmers to drill wells and buy pumps. While governments often facilitate this process through subsidies and rural electrification, large implementation departments are rare.

In consequence, there are very few government agencies that have frequent and direct contact with groundwater users. Furthermore, surface water development generally involves the diversion of flows or the construction of storage on a clearly defined stream or water body. The impact of such actions on downstream users is generally clear, at least in a conceptual sense. As a result, large bodies of customary and formal law along with the resource monitoring and enforcement systems required to implement them have developed over the long history of surface water development.

This is not generally the case with large-scale groundwater development. It is a recent phenomenon and diversions have a far less directly observable impact on other users. As a result, groundwater extraction remains highly 'individualistic' and tends to occur outside the framework of established institutions for allocating, monitoring or managing the resource base. In locations such as India, tens of millions of individuals own and operate wells. Most of these wells are on private lands. The location, use and even existence of such wells is often unknown to any individual aside from the owners and their immediate community. As a result, no established institutional basis for management exists.

Role of Surface Water

Global indicators of water scarcity tend to ignore variations in the importance of irrigated agriculture for food security among countries. They also fail to account for seasonal differences in supply. For example, more than 70 percent of the total supply in India occurs in the three monsoon months of June, July and August, when most of it floods out to the sea. Moreover, countrywide data ignore regional differences in water supply and withdrawal within the country, an example of the aggregation fallacy. Regardless of these caveats, most observers conclude that many countries do not have a surplus of water available for irrigation.

In fact, many countries do not have sufficient annual water withdrawal to irrigate their potential gross irrigated area even at high basin-irrigation efficiencies. Basin-irrigation efficiency includes all reuse of drainage water and is considerably higher than system-irrigation efficiency where drainage flow from one system is used for irrigation again downstream in another system. Most analyses indicate that, whatever the water scarcity indicator used, more than half of the world's population lives in countries with varying degrees of water scarcity. This scarcity can be physical (there is no more water), economic (the country cannot afford to develop additional water resources) or caused by a lack of social adaptive capacity. Examples of adaptive capacity are the ability to produce more value per unit water consumed, and imports 'virtual water', which is the water used to produce the crops obtained on the world market.

There is concern that more people may be affected by food insecurity as a result of water scarcity. Competition for the same resources combined with the increasing trend of water pollution exacerbates this problem. Moreover, the largely unknown impact of climate change may make water scarcity in some countries more severe. Several studies suggest that rice yields are likely to increase in the higher latitudes and decrease in the lower latitudes under future climates. It is likely that the poorest countries (and the poorest people within them) will suffer disproportionately as they are less able to adapt to the changing conditions. Future projections by the International Food Policy Research Institute (IFPRI) indicate that water withdrawals will rise by some 22 percent between 1995 and 2025. Projected withdrawals in developing countries will increase by 27 percent in the 30-year period, compared with 11 percent in developed countries. Only a very small increase in irrigated area is expected, which will be more than offset by increases in river basin efficiency.

Investments in Irrigation Infrastructure

An accurate global view of irrigation investment trends does not exist, but certain proxies can be used to indicate such trends. For example, there has been a sharp decline in World Bank lending for new irrigation schemes. Funding for new irrigation construction has largely stopped and the emphasis is on the sustainability and efficiency of existing systems. Irrigation and drainage is still one of the core investment activities of the World Bank's rural portfolio, but now mainly in support of rehabilitation and the devolution of responsibilities to water user associations (WUAs). However, the number of irrigation and drainage projects is expected to decrease to well below what it was in the 1980s.

Investments in irrigation systems are perceived to have failed to address the changing needs of irrigation services because rehabilitation of existing systems was mostly carried out to restore original project objectives. This type of rehabilitation is often inappropriate as it tends to ignore desirable changes in cropping patterns and irrigation techniques and thus allows low water-productivity practices to continue. Cost and time overruns in irrigation projects as well as public opposition to large dam building projects have further eroded the confidence of funding agents in irrigation investments.

Considering the negative aspects of irrigated agriculture (e.g. salinity, waterlogging, health hazards, and groundwater exploitation), it can no longer be taken for granted that irrigation is a protected, preferred practice and that its negative externalities will be accepted unconditionally. Nevertheless, irrigation development and dam building must continue, even if only to update existing facilities and to replace dams and reservoirs that have lost most of their storage capacity as a result of sedimentation. The loss of effective capacity of Mediterranean dams is currently between 0.5 and 1 percent/year with some as high as 3 percent. In Morocco, the reduction in regulating capacity attributable to reservoir silt-up is equivalent to a loss of irrigation potential of 6 000-8 000 ha/year. Improved erosion control in catchment areas may eventually prolong the life span of reservoirs and dams.

A reduced level of investment in irrigation may not be all bad. In the past, the construction of many irrigation systems was supply driven as part of internationally funded development aid without significant input from the future users of the scheme and sometimes against their expressed wishes. Irrigation potential was and still is seen as an important indicator for evaluating future irrigation development. This parameter expresses how much a country's irrigated area could be expanded according to land use and water availability criteria. Hence, its value changes over time depending on the country's economy and competition for water.

However, this notion of irrigation potential has often been used as the sole criterion in setting a country's agricultural and water resource policies, without a parallel analysis of economic, social, institutional and environmental constraints and without a thorough market analysis. The failure of some irrigation schemes can be attributed to a narrow focus on irrigation infrastructure and water distribution combined with an insufficient focus on the productivity of the agricultural systems and their responsiveness to agricultural markets.

The public policy of supply-driven irrigation development, adhered to by many governments and donor agencies, may be justified because of the observed importance of the role of irrigated agriculture in food security. However, the role of the private sector in irrigation development is often underestimated or ignored. The many investment decisions of smallholders and commercial farmers could exceed public investment, e.g. in Zambia and India. In particular, where there is a comparative advantage in irrigated production to service local and international markets, which may be for vegetables and cut flowers rather than traditional food crops, significant private irrigation investments appear to follow.

Role of Irrigation in Rural Development

Since 1960, growth in average cereal yields has largely kept pace with the increase in world population. It is widely assumed that it will continue to do so until the population begins to stabilise. Most of the increase in grain production has been the result of yield increases rather than expansion in cropped area. Projections by FAO, the IFPRI and the World Bank assume that the further increases in cereal production will come from continuous increases in yield. However, trends in yield data collected by FAO indicate that the average world cereal yield would have to reach at least 4 tonnes/ha for a world population of 8 000 million people from its present level of about 3 tonnes/ha. At present, all the developed countries taken together have not achieved an average cereal yield of 4 tonnes/ha. This is the extent of the challenge.

The contribution from irrigated agriculture to achieving this goal will be critical as irrigation provides a powerful management tool against the vagaries of rainfall. Irrigation also makes it economically attractive to grow high-yielding crops and to apply the adequate plant nutrition and pest control required in order to obtain the full potential of these modern varieties. According to the IFPRI, while food production will increase much faster in developing countries than in developed countries, it will not keep pace with demand, and food imports will need to increase.

In 1999/2000, developing countries produced 1 030 million tonnes of grain, i.e. 55 percent of world production, and accounted for 61 percent of world grain consumption. To bridge the gap between demand and production, developing countries imported 231 million tonnes of grain, equivalent to 72 percent of worldwide imports. These statistics illustrate that developing countries play a major role in the international agricultural trade and that they are highly susceptible to changes in the world agricultural market in terms of food security. For the poorest countries, an increase in domestic agricultural production is key to improving food security. This explains why expectations about the food security role of irrigated agriculture remain high.

Agricultural development based on water conservation and irrigation is often considered a promising avenue for poverty alleviation in rural areas. For example, the availability of water for a small domestic garden plot, usually managed by women, can make a significant difference to household nutrition and thus contribute to improved livelihoods. Water harvesting may make this possible.

However, this effect is small scale and irrigated agriculture with its higher crop yields is expected to have greater impact on the incidence of poverty and malnutrition. This effect is expected regardless of whether the irrigation project is small or large scale. However, recent studies have shown that poverty alleviation as a result of irrigation development requires that the project be geared towards the needs of the poor. This includes access to training in the technical aspects of irrigation but also in community organisation and marketing.

One of the recurrent problems is the lack of access to credit, capital or land. Even microcredits have no grace period; repayments typically have to start after a few weeks. This makes them of little use for the purchase of cheap technology, such as treadle pumps and microdrip systems. It has been argued that these technologies are profitable within a short period and do not require a subsidised price for poor people or specific poverty alleviation measures. The problem of credit is not specific to irrigation development and needs to be addressed in a more general sense for successful rural development of poor regions.

Expanding irrigated areas, increasing the control of water and applying high-yield technology in irrigated agriculture have given rise to large increases in farm income, especially in Asia. However, this increase has been disproportionately in the hands of the larger peasant farmers. They are not the poorest of the poor, but their increased expenditure pattern has driven increased employment of those who are the poorest of the poor. The latter have little or no land and they benefit little even from agricultural production programmes directed most closely to them. However, they benefit from lower food prices, increased wages and growth in demand for rural non-farm goods and services. By contrast, the capital- and import-intensive consumption patterns of large-scale farmers, and especially absentee farmers, contribute much less to poverty reduction. This is more typical for some Latin American countries than for Asia and Africa.

Cost recovery from poor farmers for the operation and maintenance (O&M) of irrigation systems is controversial. Subsidising these services and providing irrigation water far below cost is financially unsustainable. Stepped tariffs in which the basic need is provided free to poor people may work in the case of drinking-water but is difficult to implement for irrigation water. Monitoring the efficiency of water use in agriculture for many small farmers each using a small amount of water is expensive, but providing irrigation water below cost contributes to wasting of water.

In developing countries, agriculture generally produces many non-tradable goods, such as food crops of lower quality and goods with unusually high transaction costs. This aspect gives agriculture a prominent role in poverty reduction. It also buffers the national economy from shocks to international markets in agricultural commodities. For the rural poor in low-income countries, increased employment opportunities allow them to escape from poverty and hunger. Because they generally have few skills, the poor are more likely to find employment in the production of goods and services that cannot be marketed on the international market.

Examples of this type. of employment include maintenance of irrigation and drainage structures, watershed management, and afforestation, and where there is a sizeable storage reservoir, employment could be found in fisheries, ecotourism and navigation. Thus, increased employment and, hence, poverty reduction depend on increased domestic demand for these non-tradable, non-farm goods and services. Agriculture is the principle source of such demand and so it is only with rising farm incomes that poverty can be reduced and food security increased.

Hence, investments in irrigation development may achieve additional goals such as enhancing economic growth and poverty alleviation in rural areas. Nonetheless, the question may be asked whether investments in other parts of the infrastructure are not more likely to achieve these goals. For example, the steady decline in poverty in India from the mid-1960s to the early 1980s was strongly associated with agricultural growth, particularly the green revolution, which coincided with massive investments in agriculture and rural infrastructure.

According to IFPRI studies in India, the impact of additional irrigation investments on poverty reduction ranked third after rural roads, and agricultural research and extension. Additional government spending on irrigation had a significant impact on productivity growth, but no discernible impact on poverty reduction. While spending on irrigation and power has been essential in the past for sustaining agricultural growth, the levels of irrigation may now be such that it may be more important to maintain rather than increase the systems. The IFPRI studies have also indicated that the marginal returns to several infrastructural investments in India are now higher in many rainfed areas. They also have a potentially greater impact on reducing rural poverty.

A global analysis of the link between farming systems and poverty indicates that the prospects for reducing agricultural poverty in the Near East and North Africa are good. However, for the region as a whole, exit from agriculture is the best household strategy for poverty reduction, followed by increased off-farm income. The study indicates that the priority roles of the State are to support the development of vital infrastructure, such as roads, water supplies, services and power supply, and to regulate resource use and pricing for water and power. By comparison, in South Asia, measures that support households in small-farm diversification and also for growth in employment opportunities in the off-farm economy were found to be the most likely to contribute to poverty reduction.

When weighing the pros and cons of new irrigation investments against the benefits of other investments, all additional potential benefits of irrigation, such as health benefits resulting from better nutrition (i.e. more calories and a more balanced diet) and greater rural employment, should be taken into account. Many of the irrigation benefits are site-specific and generalisations cannot be made. Moreover, without proper techniques for monitoring the physical performance of irrigation systems, it is impossible to assess the potential benefits that may accrue from further investments to improve them.

Notwithstanding these caveats, the fundamental question concerning the economic utility of further investments in irrigation development as a means towards rural development and poverty alleviation is an important one. At least two conclusions can be drawn from the discussion of the role of water in sustainable food production, poverty alleviation and rural development. The first is that donor agencies and governments have a difficult choice to make when investing for poverty reduction and rural development. The choice is not automatically for agriculture or water. The second conclusion is that the right set of government policies can make a large difference in food production, poverty reduction and rural development.

Agricultural Water Productivity

Productivity is a ratio between a unit of output and a unit of input. Here, the term water productivity is used exclusively to denote the amount or value of product over volume or value of water depleted or diverted. The value of the product might be expressed in different terms (biomass, grain, money). For example, the so-called 'crop per drop' approach focuses on the amount of product per unit of water. Another approach considers differences in the nutritional values of different crops, or that the same quantity of one crop feeds more people than the same quantity of another crop. When speaking of food security, it is important to account for such criteria.

Another concern is how to express the social benefit of agricultural water productivity. All the options that have been suggested can be summarised by the phrases 'nutrient per drop', 'capita per drop', 'jobs per drop', and 'sustainable livelihoods per drop'. There is no unique definition of productivity and the value considered for the numerator might depend on the focus as well as the availability of data. However, water productivity defined as kilogram per drop is a useful concept when comparing the productivity of water in different parts of the same system or river basin and also when comparing the productivity of water in agriculture with other possible uses of water.

Crop water production is governed only by transpiration. As it is difficult to separate transpiration from evaporation from the soil surface between the plants (which does not contribute directly to crop production), defining crop water productivity using evapotranspiration rather than transpiration makes practical sense at field and system level. In irrigated agriculture in saline areas, the leaching requirement, i.e. the amount of water that needs to percolate to maintain rootzone salinity at a satisfactory level, should also be included together with evapotranspiration in the amount of water that is necessarily depleted during plant growth. Other non-productive but beneficial uses could be included. Examples are evapotranspiration by windbreaks, cover crops, and the water used in wetting seedbeds to enhance germination.

The question of considering water losses from seepage and field percolation as consumption does not receive a unique response. If this water is of no use downstream or if it generates further pollution such as that resulting from geological salt leaching, then it must be accounted for as consumption. Solutions to minimise these losses, such as canal lining or water improvement application, then have a positive effect on productivity. However, from a broader environmental point of view, it can be important to consider the impact of the outflow of an irrigation system on the overall productivity of an ecosystem.

As with the numerator, the choice of the denominator (which drops to be included) should depend on the scale, the point of view and the focus. At basin level, the choice might be between water diverted from the source and the same minus water restored, whereas at field level one might consider useful rain, irrigation water and supplemental irrigation.

Spatial Variability

Reported data on water productivity with respect to evapotranspiration (WPET) show considerable variation, e.g. wheat 0.6-1.9 kg/m^3, maize 1.2-2.3 kg/m^3, rice 0.5-1.1 kg/m^3, forage sorghum 7-8 kg/m^3 and potato tubers 6.2-11.6 kg/m^3, with incidental outliers obtained under experimental conditions. Data on field-level water productivity per unit of water applied (WPirrig), as reported in the literature, are lower than WPET and vary over an even wider range. For example, grain WPirrig for rice varied from 0.05 to 0.6 kg/m^3, for sorghum from 0.05 to 0.3 kg/m^3 and for maize from 0.2 to 0.8 kg/m^3.

The variability occurs because data were collected in different environments and under different crop management conditions. These affected the yield and the amount of water supplied. Furthermore, it is often difficult to determine the real crop yield over a large area, e.g. the size of a large irrigation system. When asked for yield figures, individual farmers are likely to give a figure that depends on the situation. For a loan application, they may overstate the yield, whereas for payment of a debt or a tariff, they will probably understate the yield obtained. Vegetable yields of vegetables may change every day, and unless good records are kept, no one will know exactly how much was harvested during the total harvest period. Yields expressed in monetary terms are more doubtful as prices on the local market may fluctuate considerably over time.

Nevertheless, water productivity data across scales are useful in assessing whether water drained from upstream is reused effectively downstream. However, there are few reliable data on water productivity at different scale levels within the same system. A study using remote sensing and GIS technologies assessed crop WPET at various irrigation system scales in the Indus Basin in Pakistan.

Crop water productivity was found to vary significantly at the scale of small canal command areas. When water productivity was aggregated for canal command areas, the highest water productivity values decreased gradually. Their variability also decreased until at a scale of about 6 million ha water productivity tended to a low value of about 0.6 kg/m^3. This arose because at the larger scale, canal commands with less fertile or saline soils and with less canal water and poorer quality groundwater were included in the average.

Increase of Water Productivity

Despite concerns about the technical inefficiency of water use in agriculture, water productivity increased by at least 100 percent between 1961 and 2001. The major factor behind this growth has been yield increase. For many crops, the yield increase has occurred without increased water consumption, and sometimes with even less water given the increase in the harvesting index. Example of crops for which water consumption experienced little if any variation during these years are rice (mostly irrigated) and wheat (mostly rainfed), for which the recorded increases worldwide amount to 100 and 160 percent respectively.

At the global level, the increase in water consumption for agriculture in the past 40 years has been 800 km^3 while world population has doubled to 6 000 million. Considering that the arable rainfed area has not increased, one can conclude that with an additional 800 km^3 of water the world has been able to feed an additional 3 000 million people. This gives a rough estimate of 0.720 m^3/d/capita. This figure is low compared to the estimated global average for 2000 of 2.4 m^3/d/capita, which includes water for food at field level not including water losses. This is a good indicator of the significant productivity gain recorded in agriculture; a gain that has enabled the world to accommodate the doubling of the population and also increase intake. As a whole, one can estimate that the water needs for food per capita halved between 1961 and 2001 from about 6 m^3/d to less than 3 m^3/d.

The importance of water needs for food makes any small relative gain in this sector equivalent to a significant gain for other uses. For example, given the water needs for capita in 2000, a 1-percent increase in water productivity in food production generates a potential of water use of 24 litres/d/capita. In order to produce the equivalent of the domestic water supply, a gain of 10 percent in agricultural water productivity would be required, which is a matter of years. Therefore, it can be argued that investing in agriculture and in agricultural water is the best avenue for freeing water for other purposes.

However, future agricultural gains will need to be split into several components:

(i) compensation for the reduction of agricultural production areas as a result of urban encroachment, soil degradation, and the depletion of water resource availability or access (groundwater);

(ii) increased water access for the rural poor and vulnerable groups;

(iii) generation of wealthier farming systems; and

(iv) freezing water for other uses including the environment.

Principles for Improving Water Productivity

The key principles for improving water productivity at field, farm and basin level, which apply regardless of whether the crop is grown under rainfed or irrigated conditions, are:

(i) increase the marketable yield of the crop for each unit of water transpired by it;

(ii) reduce all outflows (e.g. drainage, seepage and percolation), including evaporative outflows other than the crop stomatal transpiration; and

(iii) increase the effective use of rainfall, stored water, and water of marginal quality.

The first principle relates to the need to increase crop yields or values. The second one aims to decrease all 'losses' except crop transpiration. Its phrasing does not imply that it will be impossible to increase water productivity by reducing stomatal transpiration. It is conceivable that plant breeding may find ways to overcome this constraint. The third principle aims at making use of alternative water resources. The second and third principles should be considered parts of basinwide integrated water resource management (IWRM) for water productivity improvement. IWRM recognises the essential role of institutions and policies in ensuring that upstream interventions are not made at the expense of downstream water users.

These three principles apply at all scales, from plant to field and agro-ecological levels. However, options and practices associated with these principles require different approaches and technologies at different spatial scales.

Enhancing Water Productivity at Plant Level

Plant-level options rely mainly on germplasm improvements, e.g. improving seedling vigour, increasing rooting depth, increasing the harvest index (the marketable part of the plant as part of its total biomass), and enhancing photosynthetic efficiency. The most significant improvements in yield stability have usually resulted from breeding programmes to develop an appropriate growing cycle such that the duration of the vegetative and reproductive periods are well matched with the expected water supply or with the absence of crop hazards.

Planting, flowering and maturation dates are important in matching the period of maximum crop growth with the time when the saturation vapour pressure deficit is low. The periods of maximum crop growth may be optimised by means of breeding technology. Improved varieties with a deeper rooting system contribute to drought avoidance and the effective use of water stored in the soil profile. Drought escape and increasing drought tolerance are also important strategies for increasing water productivity.

Daylength-insensitive varieties of short to medium duration enabled crops, such as wheat, rice and maize varieties developed as part of the green revolution, to increase water productivity by escaping late-season drought that adversely affects flowering and grain development. The modern rice varieties have about a threefold increase in water productivity compared with traditional varieties. Progress in extending these achievements to other crops has been considerable and will probably accelerate following the recent identification of the underlying genes. Genetic engineering, if properly integrated in breeding programmes and applied in a safe manner, can further contribute to the development of drought tolerant varieties and to increasing the water use efficiency.

Raising Water Productivity

Improved practices at field level relate to changes in crop, soil and water management. They include: selecting appropriate crops and cultivars; planting methods (e.g. on raised beds); minimum tillage; timely irrigation to synchronise water application with the most sensitive growing periods; nutrient management; drip irrigation; and improved drainage for water table control.

Water depletion occurs when water evaporates from moist soil, from puddles between rows and before crop establishment. All cultural and agronomic practices that reduce these losses, such as different row spacings and the application of mulches, improve water productivity. The irrigation method also affects these evaporative losses.

Drip irrigation causes much less soil wetting than sprinkler irrigation. The significance of soil improvement in enhancing water productivity is often ignored. However, integrated crop and resource management practices, such as improved nutrient management, can increase water productivity by raising the yield proportionally more than it increases evapotranspiration. This principle applies to both irrigated and rainfed agriculture. Integrated weed and integrated pest management have also contributed effectively to yield increases.

One of the field-level methods for increasing water productivity is deficit irrigation, where deliberately less water is applied than that required to meet the full crop water demand. The prescribed water deficit should result in a small yield reduction that is less than the concomitant reduction in transpiration. Therefore, it causes a gain in water productivity per unit of water transpired. In addition, it could lower production costs if one or more irrigations could be eliminated.

For deficit irrigation to be successful, farmers need to know the deficit that can be allowed at each of the growth stages and the level of water stress that already exists in the rootzone. Most importantly, they need to have control over the timing and amount of irrigations. Deficit irrigation carries considerable risk for the farmers where water supplies are uncertain, as is the case with rainfall or unreliable irrigation supplies.

Where water availability falls below a certain level, the value of the crop can fall to zero, either because the crop dies or because the product is of such low quality as to be unmarketable. When water is scarce, farmers could reduce the irrigation as appropriate to maximise returns to water if they have control over the timing and amount of irrigations. This degree of flexibility is usually the case with sprinkler and drip irrigation, and also with pumped groundwater if the farmer owns the pump. A totally flexible delivery system for surface irrigation in large irrigation systems is expensive because of the required overcapacity in the conveyance system.

The trade-off between reduced yield and higher water productivity needs to be quantified in economic terms before recommending deficit irrigation (and other water-saving irrigations in rice production). The often cited low water productivity per unit of water supply in rice cultivation derives from considering as losses the percolation resulting from the standing water layer on the field surface. However, this water is often recycled, and rice water productivity generally compares well with that of a dry cereal. Nevertheless, water-saving irrigation techniques such as saturated soil culture and alternate wetting and drying can reduce the unproductive water outflows drastically and increase water productivity. These techniques generally lead to some yield decline in the current lowland rice high-yielding varieties.

However, some experiments are reporting substantial yield increases for local varieties using a technique called system rice intensification (SRI), a technique which originated in Madagascar. Here again there is no unique response; the fit with local resources and capacity is the most important feature to account for. Without anticipating results of current investigations in many countries, it seems that the potential of the SRI technique for the poor to increase the productivity of scarce land and water is significant provided that enough family labour is available.

Other approaches are being researched as part of efforts to increase water productivity without sacrificing yield. One of these is to develop so-called aerobic rice systems that allow rice cultivation in non-flooded conditions. The development of these new rice varieties is essential if rice is to be grown like other irrigated upland crops and the deep percolation associated with paddy rice is to be avoided.

Water-related problems in rainfed agriculture are often related to large spatial and temporal rainfall variability rather than low cumulative volumes of rainfall. The overall result of rainfall unpredictability is a high risk for meteorological droughts and intraseasonal dry spells. Bridging crop water deficits during dry spells through supplementary irrigation stabilises production and increases both production and water productivity dramatically if water is applied at the moisture-sensitive stages of plant growth.

Water harvesting for agriculture involves a storage reservoir, while in runoff farming the collected runoff is applied directly to the cultivated area. Either way, the investments in the construction of the ditches that take the runoff to the storage reservoir and of the reservoir itself are relatively small. Maintaining these structures may be more difficult if heavy rains periodically wash them away.

Many factors affect the success of rainwater harvesting. These include: the method used for runoff collection and storage; the topography; the soil characteristics (especially the infiltration rate); the choice of crop to be planted; fertiliser availability; and the effectiveness of the soil crust in the catchment area. However, probably more important than any of these physical parameters is the involvement of the beneficiaries in the design and implementation of the water harvesting structures.

Socio-economic assessments of water harvesting and supplementary irrigation are rare. It is recognised that sustainable increases in water productivity by water harvesting can only be achieved through a combination of farmer training, water conservation, supplementary irrigation, better crop selection, improved agronomic practices, and political and institutional interventions. Planning (and economic assessment) should consider explicitly the short-term effect and longer-term implications of hydrological changes brought about by water harvesting on downstream water users.

Accounting for Water Productivity

Changing the focus from the field level to system and river-basin level changes the relative importance of the various water management processes. At the larger scale, the effect of agriculture on other water users, human health and the environment becomes at least as important as production issues. Options for improving water productivity at the agro-ecological or river-basin level are found in: better land-use planning; better use of medium-term weather forecasts; improved irrigation scheduling to account for rainfall variability; and conjunctive management of various sources of water, including water of poorer quality where appropriate. Therefore, integrating germplasm improvement and resource management is crucial in the enhancement of water productivity at the field scale and above.

Gains in water productivity are possible by providing more reliable irrigation supplies, e.g. through precision technology and the introduction of on-demand delivery of irrigation supplies. However, an increase in water productivity may or may not result in greater economic or social benefits. The social benefits represent the benefits to society resulting from the water-productivity-enhancing interventions. Water in the rural areas of developing countries has many uses. Thus, water is both a public and a social good, a fact that complicates value calculations. These many uses of water include: the production of timber, firewood and fibre; and raising fish and livestock. Non-agricultural uses of water include domestic (drinking and bathing) and environmental uses.

An IWMI study of an irrigation system in Kirindi Oya in southern Sri Lanka illustrates the importance of the multiple roles of water in agriculture. The study found that at system level crops consumed only 23 percent of the total water supply, including both rainfall and external irrigation water. Of the remainder, 8 percent was used for grazing land, 6 percent evaporated from the reservoir, 16 percent was lost to the sea, 3 percent drained into lagoons, while as much as 44 percent of the water supply went to perennial vegetation that had developed since the construction of the scheme. This perennial vegetation was there because of irrigation seepage and recharge of the shallow groundwater. Tree growth is important to the people living in the area as it provides them with shade and thus improves their environment. In this project, as well as in many places in southern India, it also provides income from coconut and materials for construction (beams and ropes). Other trees are important for additional nutritional values (fruits) and some are crucial for their medicinal properties. A changeover to total control of irrigation outflow in order to increase water productivity would cause the collapse of the entire local agroforestry system.

Another example of the economic and social benefits of agroforestry is a project located along the Niger River in Mali. In this project, trees were planted on the bunds of rice fields, and also in the middle of the rice fields without affecting rice yields adversely. In this remote arid part of Mali, the value of the wooden poles of seven-year-old eucalyptus trees was so high that the farmers could pay for the O&M of the irrigation system from the sale of the trees. In another irrigation system, in southwest Burkina Faso, oil-palm and fruit trees were combined successfully with irrigated crops (mainly maize, groundnuts and industrial tomatoes). Trees were planted on ridges or on the boundaries between parcels. On the sandy, percolating soils of the irrigation system, the trees produced an important amount of complementary food and income, while the impact on the main crop was minimal.

These examples point out that not all measures to increase water productivity are appropriate in all circumstances. It is essential to consider the various uses of water in agriculture before measures are introduced that would increase water productivity at the expense of other benefits from the same source of water, especially those benefits that accrue to the local poor and landless people.

Promoting Water Productivity Gains

Using price policies to promote the economic productivity of water requires significant government intervention in order to ensure that equity of access to water and public-good issues are covered adequately. Some studies in the Indian subcontinent and elsewhere have suggested that the price for water that would be required to affect demand substantially would be about ten times the charge required to cover the O&M of the irrigation system.

A charge sufficient to cover O&M would have a minimal effect on water demand. Moreover, introducing volumetric charges for irrigation water is difficult and involves considerable expense for the installation of measuring structures and for fraud prevention. Last, in most rice-based systems in Asia, volumetric charging at individual user level or even group level is unsuitable given the permanent overflow and recycling water flows throughout the command area.

The groundwater market in India illustrates the perhaps unintended impact of government policies on the availability of water to farmers and others. Farmers in Gujarat paid about four times as much for pumped groundwater compared with farmers in Punjab and Uttar Pradesh. This difference was attributed to:

(i) differences in the way farmers were charged for the electric power to run their pumps (flat rate versus per unit consumed);

(ii) the tubewell spacing policy in Gujarat that gave each tubewell owner a monopoly over some 203 ha; and

(iii) the scarcity of public tubewells in Gujarat, which also reduced competition among groundwater suppliers.

The high prices for tubewell water in Gujarat discriminated against small and poor farmers. However, some simple changes in water policies for power pricing, tubewell spacing and public tubewells could transform groundwater markets in Gujarat into powerful instruments for small-farmer development.

Aiming for the highest economic productivity of water in agriculture may conflict with the political desire for national food security. More often than not, the economic productivity of water in growing staple crops is less than that for growing vegetables or flowers for export markets. Crop substitution involves switching high water-consuming crops for less water-consuming crops or for crops with higher economic productivity. The approach provides a strategy for increasing crop water productivity at the agro-ecological system level as well as at the global level.

Policies and incentives are important in the adoption of changes from traditional agronomic and cultural practices. However, it is necessary to identify the types of policies and incentives that will work best. Experience with conservation agriculture indicates that the short-term interests of the farmers often differ from the long-term interests of society and that the financial benefits that accrue from changes in cultural practices often take a long time to materialise. In addition, although there are large differences between individual farms, external factors also play a role, e.g. the transmission of information (via policy-related activities and social processes).

Of particular importance is the fact that the inconsistent and sometimes contradictory results from studies on the adoption of new practices suggests that the decision-making process is highly variable. This decision-making process needs to be understood more fully as it will affect the lead time from study to field practice. This lead time is often unacceptably long considering the urgent character of water-scarcity problems. Experience from participatory research and extension could help reduce this lead time.

Risk Management in Agricultural Water Use

Vulnerability to drought varies from country to country. It depends inter alia on the stage of development. Economies in the early stages of transition from subsistence farming to a more modern and productive farm economy are particularly vulnerable. This applies to much of rainfed agriculture. Rainfall patterns over Africa have not changed significantly in the past century. In particular, the Sahel, the Horn of Africa and the countries around the Kalahari Desert are characterised by high interannual and intraseasonal rainfall variability. Good and bad years do not occur at random but tend to be grouped. This has important implications for food security as food and water must be stored over a period of several poor years.

Risk is defined as the product of hazard and vulnerability. In other words, it relates to the probability of a damaging event, such as drought, and the foreseeable consequences of such an event. In terms of agriculture, the most common risk is drought. On a global scale, this risk is much greater than that of cyclones, floods and storms. However, on a regional rather than global scale, there are areas where the risk of flooding exceeds that of drought. Drought represents one of the most important natural triggers for malnutrition and famine. Drought events can be addressed at the parcel level by several management decisions, at the watershed level and at the country level. The first decisions belong to farmers or farmers collectivities whereas decisions at watershed and country level must be taken by governments or state agencies.

Risk is also defined more simply as a loss due to a damaging event. The advantage of this definition is that it can be materialised and measured easily (e.g. loss of agricultural production, loss of income). An acceptable risk is one that individuals, businesses or governments are willing to accept in return for perceived benefits. Local governments usually define the level of acceptable risk by considering information on drought hazards and combining it with economic, social and political factors specific to the area threatened.

Conflict is an ever-present risk and one of the most common causes of food insecurity. The displacement of people and the disruption of agricultural production and food distribution leaves tens of millions of people at risk of hunger and famine. Conversely, food insecurity may lead to or exacerbate conflict. Conflict in sub-Saharan Africa resulted in losses of almost US$52 000 million in agricultural output between 1970 and 1997, a sum equivalent to 75 percent of all official development assistance received by the conflict-affected countries. Conflict combined with drought has triggered six of the seven major African famines since 1980. Early warning and response can prevent famine arising from drought and other natural disasters. In war zones, lack of security and disruption of transport and social networks impede delivery of relief aid. However, several other factors contribute to food insecurity. These include: lawlessness; lack of democracy; ethnic and religious divisions; degradation or depletion of natural resources; and population pressure.

Risk Management Strategies

Stemming from the definition of risk, there are two major ways of minimising risk, either by reducing hazard or by reducing the vulnerability. Ways of minimising hazards are few and can include: rainmaking; avoidance of hail; and watershed management to content floods. Ways of minimising vulnerability can include: development of surface (including pumping water from streams) and underground irrigation facilities; integrated management of water resources; ecosystem development and diversification; education and training of farmers; early warning systems; seasonal climate forecasting; and crop insurance.

Early warning systems and seasonal weather forecasts are increasingly available to provide timely information so that governments and international aid agencies can take the necessary measures to reduce the impact of drought. However, seasonal forecasting skills are imperfect and the forecasts are not yet available to farmers. If they were, those forecasts could help farmers in choosing less water-demanding crops, e.g. sorghum rather than maize, when a drought is predicted. In the absence of reliable information about expected seasonal rainfall, some farmers will tend to accept risk in anticipation of greater profit, while others will tend to avoid risk even if there is a potential for high profit. Such risk avoidance or acceptance is a personal and a cultural characteristic.

The historical evolution of irrigated agriculture was a response to reduce the risk of crop failure in lands that were subject to periodic droughts, such as the basins of the Euphrates and Tigris rivers. Crop practices and field management provide several means for coping with soil water management. Strategies in rainfed agriculture are based on producing more food per unit of rainfall in a durable manner by collecting the maximum amount of rainfall at community, farm and parcel levels, minimising water loss at farm and parcel levels, and using water efficiently at parcel level.

The collection of maximum rainfall may involve both state and farmer organisations (water harvesting, use of recycled water from other sectors) or farmers alone (water harvesting at the farm, runoff reduction at parcel level, early planting, fallow cropping system, etc.). Minimising water loss involves farmers (evaporation reduction by mulching or rapid crop cover, windshields, minimum tillage, weeding, etc.). The efficient use of water requires farmers' involvement (use of low water consuming crop species, adapted fertilisation to available water, disease and pest control, optimal planting and seeding, selected varieties able to accomplish their cycle within the climate growing period, etc.).

Risk may be reduced substantially, while expected profit is reduced relatively less, by choosing combinations of alternatives rather than any single alternative. For example, a farmer in a rainfed area, such as Machakos in Kenya, where on average a crop of maize may yield well in one out of four years, could choose to seed one-quarter of a field with maize every year. The reality is more complicated as both the total seasonal rainfall and its distribution during the growing season have a large effect on crop yield.

The above strategies enable improved use of the available water at the parcel level. Moreover, traditional farming aims at a stable production rather than a maximum income. Farmers achieve this objective through diversification of production and low-input practices that provide that do not entail too much investment or cash. Association between farmers, for example at the village scale or within farmer groups, can further reduce the risk of low production.

Spreading Risk

Crop insurance constitutes the most explicit risk-spreading mechanism that helps distribute the cost of weather-related events through financial institutions among other economic sectors and governments. Successful examples include crop insurance for the impact of cyclones and hailstorms. The impact of drought is greater in developing countries than in developed countries, but farmers in developing countries have at best only limited access to insurance. The cost of insurance for relatively low-value staple crops is usually unaffordable.

However, spreading risk can also lead to water sharing. Water transfers within countries have occurred for some time. Some canals were constructed for navigational reasons, others to supply drinking-water to water-scarce cities, and others for agricultural purposes or various combinations of these causes. Well-known examples include the Snowy Mountain Scheme in Australia and various aqueducts in California, the United States of America. Internationally, an extensive system of link canals between the branches of the Indus River was constructed in order to ensure equity of water access between India and Pakistan following partition in 1947.

China is developing extensive water transfer schemes linking the south of the country to the water-scarce and populous north. Funding and implementing such expensive schemes in the future may help to reduce the risk of international conflict over water. Where several countries share water resources, e.g. in the river basins of the Mekong, Nile, Euphrates and Tigris, there is a perceived risk that the combination of population growth, poverty, food insecurity and water scarcity might lead to conflicts over water. Current attempts at mediation through the establishment of river basin authorities aim to reduce these risks.

Reducing Environmental Impact of Agricultural Water Development

Most production systems, agriculture included, can cause both positive and negative side-effects, or externalities that are not accounted for in markets. Agriculture's positive and negative environmental services are unintended consequences of market activities that have an impact on people other than the producer of the effect. These by-products tend not to be priced in the market and, hence, their economic values are unknown or difficult to assess.

Consideration of all the positive externalities of agriculture is not readily possible. There are cases where the same phenomenon may be positive in certain circumstances and negative in others, or it may be valued positively by some observers and negatively by others. A positive externality may reduce a negative one, and vice versa. In addition, positive and negative externalities are often linked closely, e.g. soil salinity and improved employment opportunities in irrigated agriculture.

Moreover, positive externalities are often ignored whereas negative ones tend to be reported widely. A well-known example of a negative externality is the loss of biodiversity as a result of draining wetlands for agriculture. Such losses are accelerating as human settlement continues to impinge upon wetlands and forests.

Many agricultural systems have become efficient transformers of technologies, non-renewable inputs and finance. They can produce large amounts of food, but have substantial negative impacts on capital assets. These assets comprise not only the natural resources of soil and water per se but also nutrient cycling and fixation, soil formation, biological control, carbon sequestration and pollination. The issue raises concerns about what constitutes success in agricultural production if large yield increases come at the cost of environmental and health problems. One problem is that the benefits and costs accrue to different people and are not measured in the same units.

In the 1970s and 1980s, some people considered energy to be such a common measure. Indeed, sustainable systems are much more energy efficient than modern high-input systems. Low-input rainfed rice in Bangladesh and China can produce 1.5-2.6 kg of cereal per megajoule of energy consumed. This is some 15-25 times more efficient than irrigated rice produced in Japan and the United States of America. On average, sustainable systems produce 1.4 kg of cereal per megajoule compared with 0.26 kg/MJ in conventional systems.

Modern agricultural systems depend heavily on external inputs, largely derived from fossil fuels. In most industrialised countries, energy is cheaper than labour.

Hence, it seems rational to overuse natural resources and underuse labour. The result has been adverse, long-term effects on the environment. Although labour is cheaper than energy in many developing countries, agriculture often has negative effects on the environment. In relation to their policy implications, the environmental externalities of agriculture operate at different geographic scales, e.g. carbon sequestration (a positive externality) on a world scale, but salinisation of a watershed on a local scale (a negative externality).

Applying concepts such as the 'polluter pays' principle, cost recovery and cost sharing may prove unrealistic, impractical or politically disastrous to governments in countries where millions of people are poor and small-scale farmers are trying to make a living on marginal lands. A common concern in developing countries is how agricultural production in marginal areas can fulfil its primary function without depleting the natural resource base. For these reasons, developing appropriate technologies, assigning individual or common property rights, and the promotion of alternative employment outside the agricultural sector will be key strategies.

Salinity and Drainage Question

Much of the environmental impact of irrigated agriculture is linked to the management of water and salt balances of irrigated lands. This includes both minimising the amount of water required to remove salt from the root-zone, and minimising the land area required to store the salt temporarily or permanently. Good management has proved difficult. Although human-induced salinity problems can develop swiftly, solutions can be time consuming and expensive.

Various improvements in irrigation and agronomic practices can be introduced depending on the type of salinity and on the cause of the accumulation of salts to harmful levels in the rootzone. The fact that saline waters have been used successfully to grow crops shows that under some conditions, e.g. in Mediterranean climates with winter rains, saline water can be used for irrigation. Experience in other locations, where negative long-term effects from irrigating with saline or sodium-rich waters have been observed, indicates that more permanent interventions in the water and salt balance are generally required.

All arid-zone rivers have natural salt profiles, attributable to mobilisation of salts in the catchment area and saline seeps. An additional cause of river salinity is irrigation-induced transport of fossil salts owing to pumping from the groundwater into drains that discharge into the river.

Most of the drainage water from agricultural land in Punjab, Pakistan, is reused, either from surface drains or pumped up from shallow groundwater. In fact, in some systems in Punjab, one-half to two-thirds of the irrigation water is pumped from the groundwater. Therefore, the leached salts are returned to the land rather than disposed of in the river system or in evaporation ponds. The average salt influx for the Indus River water is estimated to be about twice the amount that flows out to sea. Hence, half of the annual salt influx remains in the land and the groundwater.

Most of the accumulation takes place in Punjab. A more extensive drainage system is needed in order to maintain a sustainable salt balance in the irrigated lands. Worldwide, only 22 percent of irrigated land has a drainage system (less than 1 percent of irrigated land has subsurface drainage). This makes it inevitable that more land will go out of production because of waterlogging and salinity. In general, those people who will lose their land are already very poor farmers.

The drainage situation in Pakistan is in sharp contrast with that in Egypt. In Egypt, subsurface drains that take the drainage water back to the river underlie a large portion of the irrigated land. The salts do not stay in the Nile Basin but are discharged into the Mediterranean Sea. During part of the year, the salt content in the lower Indus River is much lower than in the lower Nile River and more salt disposal into the Indus River could be accepted. However, during critically low-flow periods, such disposals would not be possible. The only option during such periods would be to store the drainage water temporarily for release during high-flood periods. Extending the Left Bank Outfall Drain, now operational in Sindh, into Punjab could provide a more permanent, but quite expensive, solution than the present inadequate number of evaporation ponds.

Wastewater Reuse

The reuse of municipal and industrial wastewater in irrigated agriculture is widespread. Some of the wastewater is treated before it is reused. However, much of it is not, and this causes significant environmental and health hazards. In addition, many of the treatment plants in developing countries operate below design capacity, which contributes to the discharge of untreated wastewater into irrigation and drainage canals. Concentrations of heavy metals in canal and drain sediments and in soil samples, as well as faecal coliform bacteria counts in canal and drainage water, often exceed WHO water-quality guidelines. For example, wastewater constitutes 75 percent of the total flow of the Bahr Bagar

Drain in the Eastern Delta, Egypt, effectively turning the drain into an open sewer. Soil samples in the Eastern Delta showed cadmium levels of 5 mg/kg, more than twice the natural level.

Evidence of uptake of trace elements in crops has also been reported. For example, in the Middle Delta, Egypt, cadmium levels of 1.6 mg/kg (ppm) have been found in rice. Such levels are harmful for human health, and warrant serious attention. Thus, some of the drainage water is unfit for reuse, not because of its high salt content but because of its pollution load. In addition, safe disposal of such polluted wastewater becomes a real problem. Similar cases have been reported for other countries, e.g. Pakistan and Mexico.

The challenge of managing the conjunctive use of groundwater and canal water successfully has been alluded to before. In some areas, over-abstraction of groundwater is evidenced by the rapid dropping of water-table levels. In other areas where the groundwater is too saline for agricultural production, the water table rises as a consequence of over-irrigation and seepage from irrigation canals. Much agricultural land has gone out of production as capillary rise from shallow water tables has ruined soils and poisoned crops. Reversing this process is difficult and expensive. In India, the extent of the waterlogged areas is estimated at 6 million ha. In 12 major irrigation projects with a design command area of 11 million ha, 2 million ha are reported to be waterlogged and another 1 million ha salinised.

It is estimated that salinisation alone causes 2-3 million ha/year of potentially productive agricultural land to be taken out of production. How much of this land is reclaimed (to various degrees) and then cultivated again is unknown. Pollution of groundwater by salts and residues of agrochemicals is also a common occurrence. Where slightly saline groundwater is used for irrigation, the repeated cycles of water application to the fields, seepage of the excess water and pumping it up again from the top of the aquifer increases the salt load of the groundwater. If the vertical permeability of the aquifer is restricted, only limited mixing of seepage water takes place and the top of the aquifer from which the water is pumped becomes increasingly saline. This process has been documented for several irrigation systems in Punjab, Pakistan, where conjunctive irrigation with canal water and groundwater takes place.

The poorest farmers are those most vulnerable to environmental degradation as most of them farm under difficult growing conditions. A few farmers cultivate the best lands; the vast majority of the farmers cultivate the less fertile and marginal lands. Further degradation is likely to affect the quality of the farmers' sources of drinking-water and irrigation water, the quality of their land, possibly the quantity and quality of the fish they catch, and ultimately their health.

Lack of data on water and salt balances of irrigated land and lack of knowledge on how much water (and of what minimum quality) should be committed to downstream users frustrate attempts to allocate water more equitably to users in order to improve basin-level water productivity in agriculture. The way forward to ending unsustainable practices and reducing the concentrations of salts and agrochemicals that result directly from the degradation of the soil and water resources is a consolidated and long-term effort to improve land and water management.

Generally, agriculture and rural development have not benefited from systematic environmental analysis and management. One reason for their exclusion in the past was probably the very large number of projects (large and small) that could have been referred for an assessment, which would have overwhelmed the environmental assessment agencies. Environmental impact assessment (EIA) is usually applied to physical project planning (e.g. dams, roads, pipelines and industries), but seldom to farm practices and rural development plans. As a result, inadequate planning and inappropriate land-use practices have persisted. In many areas, soil, land and water resources are used inefficiently or are degraded, while poverty and income disparities grow.

With some 30 years of experience, EIA techniques now usually consider not only biophysical impacts but also socio-economic effects on health, human migration in and out of the project area, training of local workforce, local government capacity building, etc. Government and international policies are still needed to establish appropriate legal frameworks and an institutional base for EIA for agricultural projects. These policies should include transfer of the necessary knowledge to the rural poor, e.g. through agricultural extension services, so that they can participate in the environmental assessment of agricultural water resource management and project planning.

Modernising Water Management

Modernising water management in irrigation systems can be interpreted in different ways depending on the local circumstances. One type of modernisation is the introduction of modern technologies, such as water application and distribution through pipes rather than open channels, and the use of computerised soil-water sensors to trigger water applications. However, it also comprises older capital-intensive techniques, such as canal lining and land levelling.

These techniques can only be introduced and used successfully where the farmers can be trained in their use or already possess the necessary skills. However, the technical side is only one aspect of modernisation. Equally important are fundamental changes in the institutional arrangements and regulations and improvements in the performance and efficacy of water users and their organisations.

Modernisation is a process of technical and managerial upgrading of irrigation schemes combined with institutional reforms, if required, with the objective to improve resource utilisation and water delivery service to farms. In this sense, modernisation offers a means of institutional reform with a purpose, not just reform for its own sake. It is systemic and practical without asking that all institutional elements change and it needs to be applied where irrigated agriculture has a clear comparative advantage.

Irrigation institutions need to adopt a service orientation and improve their performance in economic and environmental terms. This entails: adopting new technologies; modernising infrastructure; applying improved administrative principles and techniques; and promoting the participation of water users. Irrigation-sector institutions need to link their central task of providing irrigation services to agricultural production and to integrate their water demands and uses with other users at basin level. An enhanced appreciation of the water cascades and flows across landscapes and the circulation of groundwater within aquifers will lead to informed decisions on the use and reuse of agricultural water.

Because modernisation is usually perceived as an engineering project, its planning typically focuses on engineering and macroeconomic issues with only broad assumptions about how the delivery system and the on-farm irrigation systems are to be managed. Where the modernised system turns out to be incompatible with existing management practices or where unanticipated extensive changes in management practices are needed to take full advantage of the potential of the modernised system, then the modernisation project is likely to fail.

In addition, failure is likely if the public irrigation organisation continues as before without the involvement and participation of the water users in the system's operation and management. Only through their involvement from the beginning of a modernisation project can farmers develop a sense of ownership and be likely to care for the system. This sense of ownership should prevent several of the problems that often arise after a short time: field channels being demolished; gates being stolen or damaged; field drainage systems becoming blocked; open drains filling with sediments and weeds; and graded land becoming spoiled by bad tillage.

An important aspect of modernisation is the effect of intended plot sizes on the feasibility of the project. For example, in Navarra, Mexico, the average plot size is about 5 000 m2 while the average area owned by one farmer is 1.3 ha. It is likely that in the near future these farms will not be economically viable for two reasons:

(i) the small plot size; and

(ii) the poor condition of the irrigation systems.

The modernisation of many irrigation systems should encompass restructuring land tenure in order to ensure plot sizes that can be farmed profitably. In this system in Mexico, such a plot size is thought to be about 5 ha.

Increasing plots sizes will also allow a reduction in the investments needed to modernise the irrigation systems. In addition, farms that perform well will have the capacity to generate jobs both directly and indirectly.

Nonetheless, site-specific considerations can lead to different conclusions. In Mali, the Office du Niger, which is dedicated to rice production, allocates individual plots of at least 5 ha. This allocation of large plots to full-time, maximum-profit farmers is seen to be inconsistent with the reality of people generally pursuing diversified livelihoods, especially when trying to escape from poverty. Moreover, small plots are often used more intensively.

For example, in Zimbabwe in the early 1990s, the government changed its policy of giving farmers irrigated plots of 0.1 ha so they could supplement their income from rainfed agriculture, to giving each farmer 3-5 ha of irrigated land. The expectation was that the larger plot sizes would induce farmers to devote themselves full-time to irrigation. The policy also favoured giving the irrigated land to men, with the idea that they would be more likely to devote their energies to irrigated farming. However, productivity per unit of land and productivity per unit of water were later found to be higher on the smallholding system, and women farmers were significantly more likely than men to be oriented towards irrigation as their main source of food and income.

Large-scale irrigation development and modernisation projects tend to concentrate on the production of staple foods while ignoring the fisheries. A main issue with loss of fishery habitat, or specifically with the reclamation of wetlands for agriculture, is that once the wetland is converted to agricultural land, people can have title to it. Legal title to natural wetlands is not possible although traditional communal rights can be recognised. Fisheries are often taken for granted. Many people fail to see the benefits of wetlands and fisheries, whereas they realise that quantifiable benefits, such as agricultural production and hydropower, will flow from new development works. Examples of undervaluing inland fisheries can be found in Cambodia, Sri Lanka and Bangladesh.

For a balanced diet, rice needs to be supplemented with animal protein, and inland fisheries provide one of the cheapest and most readily available sources of such protein. However, development decisions that affect the management and use of inland waters are often made without accurate and complete knowledge of the contribution that inland fisheries make to rural livelihoods.

One lesson from many such examples of modernisation projects is that a project is not necessarily good or bad, but rather that full knowledge of the local conditions and cultures is essential for its successful implementation. The goal in modernising irrigation institutions is not only to improve water management in agriculture but also to promote integrated water resource management (IWRM), which takes into account the social, economic and environmental sustainability of all management of the water resources.

Role of Low-cost Technologies

The introduction of low-cost technologies, which could be part of the modernisation of small-scale irrigation projects, provides another example of the site-specificity of success. Inexpensive treadle pumps have been successful in some South Asian countries in extracting irrigation water from shallow aquifers. These pumps have allowed poor farmers to make good use of the available labour in their households and so increase crop production and farm income. The farmer has full control over the timing and amount of this pumped water, which given the effort involved is used sparingly. For example, the area under irrigation by one treadle pump in West Bengal, India, varies between 0.033 and 0.13 ha. Treadle pumps have also been introduced in Africa, including the urban and peri-urban areas of Ndjamena, Chad. Here, the vegetable growers rejected the pumps in favour of mechanical pumps because they could afford the cost of fuel and spare parts. However, in the more remote areas of Sarh, Chad, farmers were content with the treadle pumps and requested more of them. Treadle pumps are intrinsically pro-poor, as richer farmers would not be able to persuade household members to use them. A limitation of treadle pumps is that their use requires spare labour, which may not exist. Women are usually already overburdened. Healthy children could do the pumping, provided it does not interfere with school attendance. Therefore, it is important to assess labour availability before introducing treadle pumps.

2

On-farm Water Management

Water, or the control of water, affects most crop production activities. Sufficient water must be present in the rootzone for germination, evapotranspiration, nutrient absorption by roots, root growth, and soil microbiological and chemical processes that aid in the decomposition of organic matter and the mineralisation of nutrients. These factors are all necessary for sustaining crop growth on a particular field. At the same time, the rootzone must be sufficiently dry to ensure adequate aeration and root growth. The rootzone must also be dry enough to allow field access for performing cultural practice activities such as planting, cultivating, fertilisation, pesticide and herbicide applications, and harvesting.

Water movement through the soil is necessary in order to leach excess salts from the rootzone and so enable potential yields to be achieved. Farmers around the world are aware that farm-level land and water management practices are of prime importance for satisfying the needs of field-crop and other agricultural and horticultural ecosystems. Therefore, they endeavour to optimise the water supply of their crops within the limits of their knowledge and the farming operations practised. That is, over time, they have developed some sort of on-farm water management (OFWM) practices.

However, farmers may often be unaware that conditions for the operation of farms are changing continuously. This has consequences for farm-level water management as the improvement of OFWM is a never-ending process.

Water management can be defined as the planned development, distribution and use of water resources in accordance with predetermined objectives while respecting both the quantity and quality of the water resources. It is the specific control of all human interventions concerning surface and subterranean water. Every planning activity relating to water can be considered as water management in the broadest sense of the term. Therefore, OFWM can be defined as the manipulation of water within the borders of an individual farm, a farming plot or field. For example, in canal irrigation systems, OFWM starts at the farm gate and ends at the disposal point of the drainage water to a public watercourse, open drain or sink.

OFWM generally seeks to optimise soil-water-plant relationships in order to achieve a yield of desired products. The managers usually try to achieve this desired yield by minimising inputs and maximising outputs, so as to optimise profits. In order to accomplish this, water has to be managed skilfully through certain practices covering areas of: soil and water conservation, water application, drainage, soil amelioration, and agronomy. All this has to be done within the context of the socio-economic environment of the community and the farmer's personal situation. There are a range of tools available that enable the manager to apply these practices.

When defining OFWM, it becomes clear that the term covers not the water resources, the irrigation facilities, the laws, the farmers' institutions, the procedures and the soil and cropping systems. OFWM is concerned with how these tools and resources are used and made available to provide water for plant growth. Moreover, it encompasses all the water used for that purpose, i.e. precipitation and water applied through irriga tion. Furthermore, it includes the use of respective practices and tools to improve site conditions and to protect crops and farming property from excessive water.

In the past, extensive research work in many countries has shown that, for example, in irrigated agriculture, good OFWM practices require well-levelled fields, appropriately designed on-farm distribution systems, and a good knowledge of when to irrigate and how much water to apply. Irrigated agriculture also requires a reliable source of water, readily available when needed, and in quantities that can be distributed effectively and efficiently over the farmer's field.

In addition, soil amelioration measures such as sub-soiling and gypsum application may be necessary, as may a well-functioning drainage system. However, all this will lead to good OFWM only if the system as a whole is well managed, if the managers or farmers take the appropriate management decision at the right time, and if they make sure that the management decisions are indeed transferred into practice correctly and on time. Furthermore, part of the success of good OFWM depends on close communication and interaction among farmers and other water users of the respective catchment area as well as with the service providers and the water supply administration. This is especially the case if the measures taken on-site will be affected by off-site activities or will affect such activities.

To satisfy the objectives of water control, an adequate soil water content must be maintained in a field. This must be done in a timely manner and within constraints imposed by characteristics of the area.

— Physical —soil types, depths, and characteristics, field layout, water sources and sinks.

— Climatic- frost potential, drought potential, rainfall amounts and intensities.

— Economic — market prices, material availability, labor cost and availability.

— Social—governmental regulations, environmental concerns, safety considerations.

Water budgeting and water management evaluations are useful tools to help the manager operate the system effectively and efficiently. Water budgeting is an accounting for all water entering, leaving, and remaining on a farm or field. This includes knowing the designated use of the available water as well as the losses from the system that cannot be controlled by the manager. Water management evaluations examine the efficiency with which water is conveyed, applied to a field, and made available for crop use for optimum growth. A well-managed system will, under most circumstances, not subject a crop to either drought or flood conditions that would result in yield reductions.

Systems Components

The term "management" implies that the on-farm water management concept is one involving various decision-making processes. These decisions revolve around using water to realise optimum returns from the crop. Economic and social factors place certain restrictions on optimum returns. Components of the water management system that must be managed on a farm can be broken down into several categories.

— Irrigation

— Drainage

— Conveyance to and from fields

— Water storage

— Release of water from the confines of the farm

— Water sources and sinks

The relationships between the components are shown schematically in Figure 1.

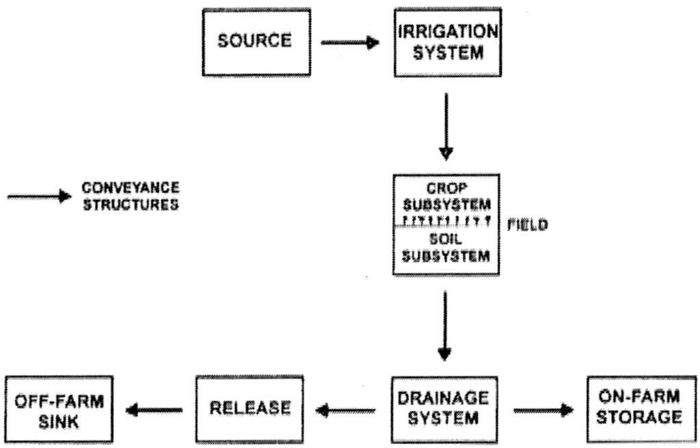

Figure 1 Relationship between systems components

To satisfy the overall objectives of the water management program, decisions must be made concerning both the types of system components to instar, as well as how, when, and how long each system should be operated. These decisions must conform to both government and other social regulations on volumes of water entering and leaving a farm.

Irrigation

The primary purpose of irrigation is to supply crops with sufficient water to satisfy evapotranspiration needs so yield reductions will not occur. The first management decision regards the type of system that will be used to perform this function. Common irrigation systems used are varied.

- — Seepage or subirrigation
- — Sprinklers — big gun, center pivot, linear move, or solid set
- — Surface — furrow or crown flooding
- — Trickle or low volume-subsurface, spray, drip emitter, bubbler, or mist

When to irrigate and how much water to apply are other concerns. Both questions can be answered by considering the evapotranspiration needs of the crop, rainfall patterns, and the need for field access. Clearly, water must be applied to the root zone before drought stress can cause irreversible damage to the crop.

When rainfall occurs in adequate amounts during major portions of the crop season, irrigation is used on a supplemental basis (in the event that rainfall is not timely or adequate). In this case, the cost of installing a sophisticated irrigation system must be weighed against the level of protection that the irrigator wishes to maintain. Irrigations must also be scheduled around the times needed to gain access to fields in order to perform cultural practice activities.

Other reasons for irrigation may also exist. Three such reasons are frost protection, pest management, and salinity control. Additional considerations are the coexistence of the irrigation system with cultural practices necessary for crop growth, the permanence of the system, and the source of water. The latter item is extremely important because quality and quantity of water may preclude the use of a particular irrigation method.

Drainage

In areas of high rainfall, adequate drainage is mandatory for crop production. In more and more regions, drainage may still be necessary to remove any excess irrigation water applied. The function of a drainage system is to remove water from a field root zone whenever necessary, and fast enough to avoid permanent root damage resulting from lack of oxygen in the soil.

Drainage systems must be compatible with crops grown, field layouts, and cultural practices such as crop rotation and cultivation. System choices include open ditches, tile drains, mole drains, and land forming for increased surface runoff. The four methods can be used separately or in different combinations. The system design must be capable of removing excess water at a high enough rate to ensure a level of protection against flood damage to crops.

Drainage system choices and operational decisions are limited by timely access to the field, economically advantageous field sizes, depths of root zones, depths of soil profiles, natural water table levels, amounts of water to be removed in a given time frame, and susceptibility of the particular crop to flooded conditions. Since drainage structures are relatively permanent, the system installed should be capable of handling the needs of all crops included in the grower's rotation.

Conveyance

Conveyance structures transport water to and from various locations on the farm, and include laterals, manifolds, submains, and mains. These delivery structures can be either open ditches or closed conduits, depending on the water source and the irrigation and drainage systems. The primary purposes of these structures are to transport water from the source (either surface or ground water) to the irrigation system, or from the field to the sink where runoff is deposited. A typical field conveyance system layout is shown in Figure 2 . A particular field does not need every component of the conveyance system. Devices such as furrows or other open ditches, sprinkler heads and drip emitters draw water from the laterals and spread it over the field surface.

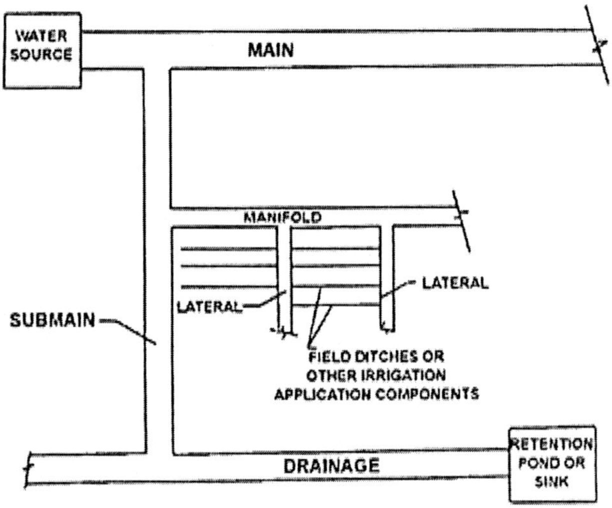

Figure 2. Typical field conveyance system layout

The conveyance structures must be large enough to handle the maximum irrigation or drainage volumes foreseen for the areas they serve, given a certain desired level of protection from flood and drought. Pressurised irrigation systems will generally require closed conduits for at least part of the conveyance or delivery system. Surface systems may use buried closed conduits to transport water in order to avoid taking land out of production. For drainage systems, open ditches, closed conduits attached to subsurface drain lines, or a combination of both may be used. If a combination is used, open ditches could carry surface runoff and subsurface drain discharge to the appropriate on- or off-farm sink.

The selection of a conveyance system should be based on land availability, access to parts of the farm (open ditches may impede ready access), volumes of water handled (burying very large conduits may be infeasible as well as expensive), rainfall amounts and intensities, water source, and type of irrigation system used. The manager or installer must then decide on how, when, and how much water will need to be routed to different parts of the farm to determine what the capacities of the structures should be. These decisions must be made to ensure that the crop is irrigated and drained in a timely fashion. To do this, a manager must know crop water needs, areas served by each structure, and the time that it takes to transport a certain volume of water in the system from one point to another.

Storage

Water can be stored on farms in various ways. The conveyance system generally provides a certain amount of storage to hold drainage amounts in excess of that removal amount planned for during design. Holding excess water in the conveyance structure may result in inadequate hydraulic head differences for drainage, or excessive seepage into a field.

The soil root zone is also a viable storage facility in which water in excess of crop needs can be held for short periods of time. The amount of water that a soil can store will depend on the soil type and the flood tolerance of the crop, both of which provide only a limited amount of storage for. a limited amount of time.

On a larger, long-term basis, on-farm reservoirs or ponds can provide water for irrigation either as primary sources or to augment the primary source in times of need. These reservoirs or ponds can hold agricultural runoff for later use, thus reducing the demand on the basin-wide system in times of drought. These ponds hold nutrient-laden water for reuse and also prevent degradation of public water supplies.

The selection of an on-farm storage facility is not always necessary. However, storage facilities provide the grower with additional insurance in the event of unduly large rainfalls or drought conditions. The use of these reservoirs is also restricted by the topography of the land and the type of irrigation system used.

For example, in areas where shallow soils overlying bedrock exist, or in areas where there are no natural catchment areas, the construction of a reservoir would be costly and only marginally beneficial since large amounts of land would have to be taken out of production. This is especially true in cases where irrigation systems require the movement of relatively large volumes of water through leaky open ditches. However, the reservoir is the most useful and powerful on-farm storage system under suitable conditions. Soils and conveyance structures can provide only temporary, small-scale relief for both irrigation and drainage needs.

Release

The final component of the on-farm water management system is the manner in which water is released from the confines of the farm. Some water will seep out of the system through the soil or percolate beneath the root zone. Water will also exit the system through evaporation from open conveyance or storage structures. However, to remove large volumes of water from a farm requires a structure capable of artificially channeling water to a point where it can either flow by gravity or by pumping to a sink outside the system. This is particularly important in areas of high rainfall.

Pumps can be used to move large amounts of water quickly. Alternatively, ditches or conduits on grades, and positioned above the water level of the sink, can perform the function adequately. Again, these structures must have the capacity to remove enough water, fast enough, to prevent crop damage given a selected level of protection.

The water manager must then make decisions for releasing water based on the drainage system capacity, crop tolerance to flood or drought, level of insurance desired in the event of needing more or less water on the farm in the future, and governmental or other special restrictions placed on water release to the public water supplies. Regulations and restrictions often involve limitations on the quality of water released to the public water supplies. The amount of nutrient-laden water that leaves a farm should be of concern to the grower, since it may be possible to reuse that water to his benefit.

Source and Sink

Two additional components of a farm water system are the water source and the sink. The source can be surface water from reservoirs, lakes, ponds, and canals, or ground water from wells. The sink can be on- or off-farm ponds, reservoirs, lakes, or canals. Once properly cleansed, excess water may also be used to recharge deep aquifers. Either gravity flow or mechanical pumping can be used to obtain water from the source and to pass runoff to the sink. In both cases, water quality must be considered an extremely important consideration.

Problems in OFWM

In addition to satisfying the needs of field-crop ecosystems, OFWM is of prime importance to soil and water conservation. However, field observations have shown that not all forms of OFWM are appropriate to achieving a sustainable land use while conserving soil and water. For example, differences in erosion due to different management practices of the same soil are often greater than the differences in erosion from different soils under the same management.

The same applies to water where differences in water use efficiency due to different management at the same site are greater than the differences in water use efficiency at different sites under the same management. Although inappropriate management practices are an old problem in OFWM, this problem now ranks higher on the agenda as the world is faced with significant population increases, as food security becomes more of an issue and as land and water resources become scarcer.

Traditional farming systems in the tropics and subtropics often included the 'good practices' of water and soil conservation. However, in many cases, increased population pressure, the introduction of new cash crops and farming systems, and the mechanisation of farming operations have led to the abandonment of these farming and conservation practices. The shift towards annual crops on steep slopes that were previously under forest, tree crops or permanent pasture has led to increased water and soil losses. Other practices such as ridge constructions, ploughing down slopes and clean weeding have created additional problems in this respect.

Problems associated with the introduction of irrigation are: waterlogging; salinisation; soil and water contamination; and a lowering of the quality of the water released through natural or artificial drainage. Overirrigation is a common fault that together with seepage from unlined canals causes the groundwater table to rise. This leads to waterlogging and salinisation especially where there is no natural drainage or where the drainage system is inadequate. In addition, yields of crops commonly grown at the site are also reduced if the soil is compacted and soil aeration and water permeability are reduced.

Another old problem is the overuse of groundwater. The new aspect of this problem is that it has become increasingly widespread. Overexploitation leads to lower ground-water levels and increasing costs of supply, and, under certain circumstances also to subsidence, landslips, seawater intrusion, etc. All these problems have been well known to agriculture for some time. They are mostly the result of inappropriate land use decisions and/or the inability to adjust to changing circumstances. They are quite often influenced significantly by non-agricultural factors such as those of a social, economic or political nature. In many cases, political factors are more important than the technical considerations usually discussed in connection with onfarm land and water management issues.

Traditional management practices of the irrigation supply and conveyance systems often contribute to high water losses. On many farms, the low irrigation efficiency is further accentuated by farmers' traditional irrigation methods and practices, inadequate land levelling, lack of a crop-specific water application, insufficient drainage, and poor maintenance of irrigation and drainage infrastructure.

Farmers are often unaware of the possibilities of applying water in a more productive way. The potential of horticultural crops with their high land, water and labour productivity is often not adequately recognised, especially by less educated and poorer farmers. Farmers generally lack technical and economic information on improved OFWM methods and techniques and on the related aspects of more productive cropping patterns and crop management. Therefore, proper training and capacity building at all levels of OFWM would be useful.

As mentioned above, these problems have been a threat to agriculture in many countries in past decades. However, these problems have now taken on an added dimension. There is increasing public demand that the development and management of land and water be ecologically sustainable as well as economically viable. As there is a close interrelationship between land use and water resources, farmers have to be aware of this interrelationship and have to adjust their OFWM efforts to address this issue.

Hence, both on-site and off-site effects have to be considered in management planning and practices. By increasing the proportion of rainfall lost due to surface runoffas a result of inappropriate land use, problems of flooding and downstream erosion arise following rain events. As less water percolates down to provide base flow for streams, the dry season flow may be lowered and even reduced to zero. Lowering of the water table can lead to the loss of well water. Decreased soil-moisture supplies result in progressively poorer vegetation. Removal of the fertile topsoil by erosion will reduce crop yields.

Deposition of coarse sand, gravel and stones removed from steep slopes onto low-lying areas decreases the agricultural potential of the soils. Sediment deposition in channels and reservoirs resulting from upstream erosion is causing major problems. Suspended sediment represents a deterioration in water quality. Fine clay particles require expensive treatment by chemical flocculation and filtration. The contamination of water resources by agriculture with harmful substances has become a widespread problem and one of increasing concern for the nonagricultural public.

Another new feature is the fact that the agriculture sector is coming under pressure to make more efficient use of water. It has been and still is criticised for being the greatest water user while having the lowest water use efficiency and lowest output per unit of water used of all sectors. In particular, irrigated agriculture, the greatest water user of all, has been accused of being responsible for inefficient water use and land degradation. In the past, agricultural research and practices dealt solely with the subject of water for the purpose of optimising water management in order to satisfy the water needs of the crop.

It is only in recent times that the effects of agricultural activities on water resources have been studied in the context of intersectoral water management and have consequently become a matter of public concern. In areas of water scarcity, the competition between different sectors of water users is attracting increasing public attention. It is also clear that future agricultural practice and research will no longer be geared exclusively to the task of optimising the water supply of crops. Rather, there is a need to curb agricultural water consumption for the benefit of other sectors, for example, by reallocating water originally designated for crop production to supplying drinking-water and domestic water to the population. The challenge is clear: OFWM has to contribute to an increase in overall water use efficiency.

Finally, it is necessary to realise that the water issues of quantity and quality are not related solely to agricultural production. They are increasingly related to urbanisation and industrialisation. The migration of water from agriculture to urban and industrial uses is underway and increasing, driven by the fact that "water flows uphill towards money". Aggregate supply economics suggest that cities will gain in the long run because of the higher prices for water that urban users are more likely to pay.

Within the next few decades, agriculture will face a number of important challenges. On the one hand, it is necessary to increase food and fibre production in order to guarantee food and clothing for the increasing world population. This contribution is urgently needed to help combat poverty while fostering economic development. On the other hand, agriculture will face increasing competition for decreasing water resources.

The situation is exacerbated by the dwindling financial resources and increasing costs for the rehabilitation of existing irrigation and drainage systems and the setting up of new ones. Moreover, agriculture needs to meet these challenges in a political and social environment that, in many countries, is highly critical of agriculture in general and of irrigated agriculture in particular. Agriculture needs to be more sparing in its use of water resources, minimise its impact on the environment and exercise continual self-restraint by means of eco-auditing.

Until now, it has been farmers who have decided what happens on agricultural land. They make rational decisions according to their own circumstances. Their decisions are influenced by: physical factors, such as soil and climate; the socio-economic features of the community; and their own personal situation. Technical advice and the assistance available may be another influencing factor. As natural resources grow scarcer, the general public sees the environment as being increasingly endangered. Hence, farmers will face the challenge of land use decisions being influenced increasingly by non-agricultural factors with pressures coming from social, economic and political quarters.

Farmers cannot ignore the fact that agricultural activities produce a significant proportion of all pollutants entering streams, lakes, estuaries and groundwater. This is especially the case in Europe, North America and Australia, but it is also increasingly true in developing countries.

Farmers need to admit that, in order to solve the water pollution problem, agricultural non-point source (NPS) pollution has to be controlled. They have to be aware that the solutions to controlling runoff will require an integrated effort by landowners, government and organisations responsible for protecting and restoring soil and water. Osmond et al. see the first step in reducing agricultural NPS pollution as being one of focusing on the primary water-quality problem within the watershed; the water-quality use impairment must be identified and the type and source of the pollutant(s) defined. Once the problem has been defined clearly and documented, the critical area, i.e. the area that contributes the majority of the pollutant to the water resource, can be identified. Land treatment, that is the installation or utilisation of good management practices (GMPs) or better GMP systems, should then be implemented on these critical areas. Good and effective OFWM adjusted to the individual circumstances is an essential part of the GMP systems. For example, in some cases it might be important for farm runoff to be kept on the farm area where there is a risk of nutrient or pesticide contamination. This will require special OFWM efforts.

In many cases, the potential for achieving benefits from improved OFWM is substantial. Water savings, yield increases, higher productivity and higher farmer incomes are achievable. Many experiments have demonstrated a positive effect of the yields for farmers. However, in many cases, questions concerning the ability and willingness of farmers to adopt the improved practices remain. In this context, the lack of financial and economic impact assessments of various interventions and the effects of improvement measures on productivity, incomes and water use efficiency remains a major shortcoming, also for policy and strategy formulation.

Water as a means for production will become increasingly scarce and expensive, making high on-farm water use efficiencies and precise water management indispensable. Furthermore, farmers may be forced to devote more attention to the sustainability of natural resources on their land and within their catchment area. This may require and prompt changes in OFWM practices. In this respect, the challenge for farmers will be to manage the water on their farms professionally in the most sustainable and profitable way. This will require detailed planning and a clear definition of the goals and the means for achieving them within the constraints given. A sound understanding of the environment and a clear commitment to sustainable OFWM practices are essential. In order to achieve all the above, it will be necessary to implement good information management and flows. In this way, it will be possible to supply the necessary information and data on which decisions should be based.

For accomplishing GMPs in OFWM, the management structure of the individual farm is quite important. This is especially the case where the labour process is becoming fragmented in the course of development. For example, in Egypt, the head of a rural household is primarily a manager, linked to other households through exchanges and the hiring of labour, rental of machinery and land, and other relationships. One of the tasks of the head of the household is to manage the labour input in agriculture. The head of the household is also usually the only household member who follows the crop through the crop cycle. The other household members and the hired labour do most of the physical work and often lack comprehensive knowledge of crop and water management. The separation of management and farm labour causes a concentration of knowledge in the person of the manager. Hence, a de-skilling of the farm labour takes place, a fact which needs to be considered when developing and implementing strategies for the introduction of advanced OFWM practices.

With generally rising income expectations and standards of living, increased agricultural yields become necessary to satisfy the higher income demand. This is why, in the course of development, land which has been regarded as fairly fertile up to this point will become marginal land, and the previously marginal land will go out of production as it may no longer guarantee the necessary yield and income levels. Agriculture and especially irrigated agriculture will concentrate increasingly on highly productive land only. More food will be produced on an even smaller area. Achieving this requires more sophisticated management of the production technologies in general, and more advanced OFWM in particular.

The implementation of a more advanced form of OFWM constitutes a major challenge for the extension service. The conventional approach of the existing extension service has often proved not to be very successful in changing farmer behaviour. This has mainly been because extension workers often have little awareness about farmers' actual needs and problems, or about the practical value of their messages with regard to the farmers' social and financial conditions. In addition, they often lack experience. New and more participatory ways for the dissemination of information in the field of OFWM seem to be needed urgently.

A participatory group extension approach is currently being introduced in Egypt's extension system. It is based on the principle of learning and doing together. It seems to be the only alternative problemsolving concept conceivable for introducing a sustainable, advanced form of OFWM. Generally, it is necessary to conduct extensive education programmes in order to update continuously the knowledge base of the extension service staff and to encourage farmers to adopt good OFWM practices.

The introduction of an advanced form of OFWM also poses many challenges for the agricultural research community. For example, as not all forms of OFWM are good, there is a strong interest in establishing criteria of what constitutes good and bad OFWM. In this sense, OFWM investigations/research/projects cannot be primarily descriptive, although detailed descriptions of the management processes may be important. Instead, OFWM investigations/research/projects have to be predictive. However, the interest lies not only in predicting what a given management will do in a certain circumstance, but more fundamentally in predicting what would occur if certain activities were adopted. The outcomes of these predictions need to be evaluated and an attempt made to rank management activities in terms of better or worse and to discern the best if possible.

3

Water Table Management

The water table or phreatic surface is the upper limit of abundant groundwater. In the vadose zone, above the water table, the interstices between particles of earth are filled by air, or by air and water. Below it, every available space is saturated with water. The definition of the water table is the surface where the pressure head is equal to atmospheric pressure. A large amount of water within a body of sand or rock below the water table is called an aquifer. A so-called "perched aquifer" (or perched water table) occurs when the descent of water percolating from above is blocked by a shelf of impermeable rock.

The porous media in which groundwaters occur are the complex geologic materials near the earth surface; hence local details of porosity and permeability are as complex as those materials. Generally, the more productive and useful aquifers are in sedimentary geologic formations, though weathered and fractured crystalline rocks yield smaller volumes of groundwater in many environments. Among the most productive groundwater environments are unconsolidated to poorly cemented alluvial materials that have accumulated as valley-filling sediments in major river valleys and geologically subsiding structural basins.

The water table is that surface beneath which all interconnected pore space in the rock is water-filled or saturated. Groundwater recharge is the process of adding new water to the groundwater body, as through infiltration of precipitation on the land surface. Groundwater discharge is the process of water escaping from the groundwater body, as when the water table intersects the land surface, allowing water to flow out from a spring.

In undeveloped regions, or areas with high amounts of precipitation, the water table roughly follows the contour of the overlying land surface, and rises and falls with rainy or dry weather. Springs and oases occur when the water table reaches the surface. Springs commonly form on hillsides, where the earth's slanting surface may "intersect" with the water table. Other, unseen springs are found under rivers and lakes, and account for the sometimes surprisingly well-preserved water levels which occur in times of mild drought. The practice of drilling wells to extract groundwater is dependent on understanding the water table. Because wells must reach the water table, its depth determines the minimum depth of a viable well, and thus the feasibility of drilling it.

Water Table Management Practices

In simple terms, water table management is the management, control, and/or regulation of soil-water conditions in the profile of agricultural soils. Essentially, excess and deficit soil-water conditions in the soil profile can be managed to provide better plant growth conditions for the production of food. Through the implementation of proper management practices and strategies, there also can be an environmental benefit. Soil-water conditions can be managed through the use of water management structures and strategies designed specifically for the given site conditions.

Water table management consists of three basic practices. These are conventional subsurface drainage, controlled drainage, and subirrigation. Each of these are illustrated below.

Subsurface Drainage

Cropland that is susceptible to seasonal or intermittent high water table conditions usually requires subsurface drainage improvement, which serves to lower the water table to a level equal to the drain depth (Figure 1). Subsurface drainage is common throughout the flat and gently rolling areas of the Midwest, as well as in other parts of the country. Subsurface drainage improves trafficability, enhances field conditions for more timely planting and harvesting operations, and helps decrease crop damage that can result from saturated soil and standing water.

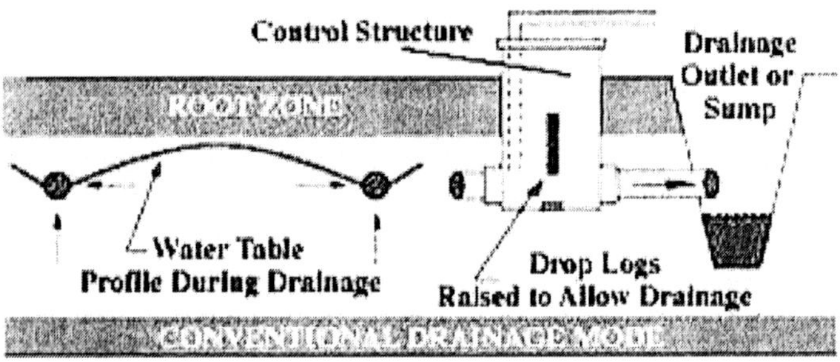

Figure 1. Conventional subsurface drainage

Two of the longest duration studies on the effect of drainage on crop yield were conducted in Ohio. Both studies document crop yield increases with subsurface drainage on poorly drained soils compared to no subsurface drainage on these soils.

Controlled Drainage

The addition of properly designed and constructed water control structures to a subsurface drainage system allows the drainage outlet to be artificially set at any level between the ground surface and the drain depth (Figure 2.) Raising the outlet after planting helps keep water available for plant use longer than does "free," uncontrolled subsurface drainage. This practice also can be used to recharge the water table between growing seasons. Most existing subsurface drainage systems can be retrofitted for controlled drainage. Controlled drainage systems require a moderate level of management so that excess soil-water conditions following heavy rainfall can be avoided.

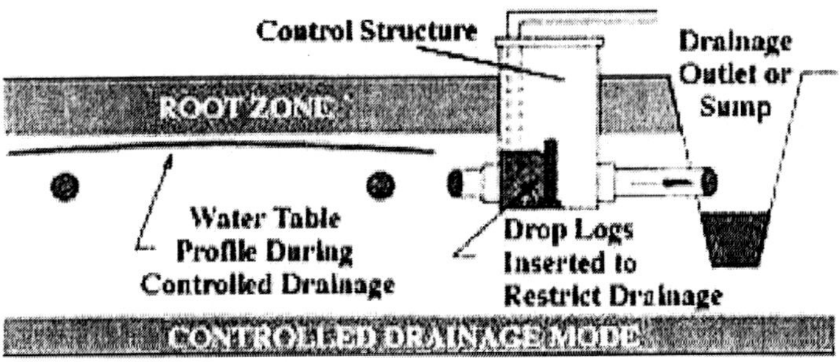

Figure 2. Controlled drainage

Based upon research conducted in North Carolina, controlled drainage may provide some reduction in nitrate losses from subsurface drained cropland, and helps to increase corn and soybean yields. Although controlled drainage has long been used in Ohio's organic soils to control subsidence and iron ochre problems, the effect of this practice on crop yields and water quality in Ohio has not been fully evaluated.

Subirrigation

With subirrigation, one system provides the drainage and irrigation requirements for the crop. Water is supplied through the subsurface drainage system using control structures to regulate the water table level in the field. Irrigation water is applied below the ground surface, thus raising and maintaining a water table at an appropriate depth in the crop root zone (Figure 3).

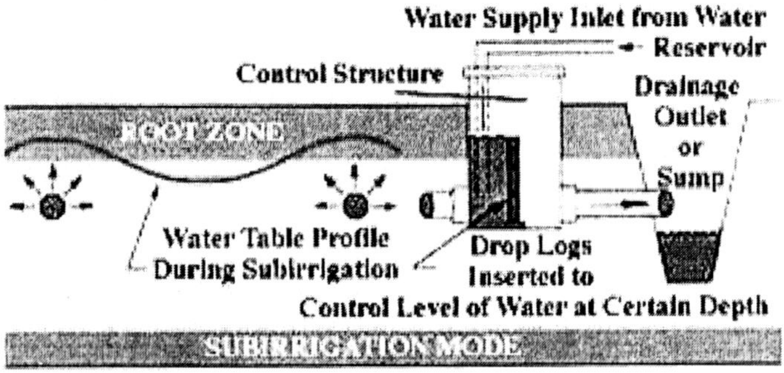

Figure 3. Subirrigation

The pumping system and water control structure can be managed to create a constant water table depth or a fluctuating water table. Some existing subsurface drainage systems may be retrofitted for subirrigation. Subirrigation systems require a high level of management to avoid excess soil wetness following rainfall.

Benefits of Subirrigation

One properly designed and managed system can be used to completely meet all the water table management requirements at a site. The drainage and irrigation components of a subirrigation system are one and the same. Installing a subirrigation system usually costs less than installing a subsurface drainage system and a surface irrigation system together on the same field.

For certain soils, subirrigation is very efficient. If the system is properly designed for the site and soil conditions, loss of water through deep seepage is negligible, and runoff of irrigation water rarely occurs. The water is always applied where the crop needs it most. Most importantly, crops respond well to subirrigation when other production management factors are not limiting. In over ten years of study in Ohio, soybean yields have been consistently over 75 bushels per acre under subirrigation and a high yield management system.

Requirements for Subirrigation

A number of factors should be considered before installing a subirrigation system. Several of the more important ones are discussed below.

Soil

In general, agricultural soils that respond well to subsurface drainage improvement tend to respond well to subirrigation. Subirrigation is usually effective in soils that have a soil layer of low permeability located below the subsurface drains. This layer helps reduce deep seepage losses. The permeability of the restrictive layer should be less than one-tenth that of the soil in the crop root zone.

Both vertical and horizontal hydraulic conductivities should be measured in the field before designing the system. High values of horizontal hydraulic conductivity creates the potential for lateral seepage. This allows for a wider drain spacing, which can reduce installation costs. However, losses from the edge of the field may be excessive under these conditions, especially if the adjoining field is drained.

Water supply

Available water from a reliable source is a very critical factor for all types of irrigation. Water is needed most during the driest parts of the growing season. Streams are often unreliable, because flow rates decrease when water demand is highest. Wells, ponds, and reservoirs are used frequently for irrigation water supply. Net irrigation water requirements in the Midwest depend on crop, location, and weather. Irrigation to meet the evapotranspiration demand for a typical Ohio growing season may require as much as 5 gallons per acre per minute per day.

Drainage

The ability to drain rapidly when rainfall occurs during subirrigation periods is critical. In addition, drainage system improvements may be necessary to adequately distribute the irrigation water throughout the field. For most soils, the subsurface drain spacing is usually closer than that required for conventional subsurface drainage alone. In general, the subirrigation system will be a more intense subsurface drainage system. Surface drainage improvements, such as land grading or field ditches, may be used to help safely and efficiently avoid ponded surface water after a rain.

Topography

Subirrigation is best suited for flat or gently sloping lands because uniform depth to the water table is much easier to maintain. A field with considerable surface undulation could result in excessive variation of the depth to the water table within the field. For this case, the field may need to be divided into zones within which the land slope variation is limited. For this type of situation, proper water table management may require a separate water control structure for each zone within the field. This will increase the cost of installation of the system, but should increase the irrigation efficiency.

Materials

The materials needed for the subirrigation system will include all the same types of materials used for a typical subsurface drainage system. In addition, the subirrigation system will require water control structures, a properly designed pumping system, and perhaps simple, water table level monitoring wells (piezometers) at several locations within each field. Water control structures are needed, at least in the main line, to maintain a uniform water table depth. Provisions for adjusting the weir setting (water level) within the water control structure must be included and should be easy to adjust and operate.

Converting from drainage to subirrigation

Subsurface drain spacings for subirrigation usually are 30% closer than those for drainage only. Retrofitting an existing subsurface drainage system for subirrigation may be possible in some cases by installing additional drains between existing lateral drains, water control structures and a pumping system. Extra mains are often required when laterals run upslope.

Management

Management is a very important aspect for water table management to be successful, and time requirements by the manager may be high. Until the operator or manager has gained much hands-on experience and is well acquainted with how the system works, daily monitoring of the water table both over and between the drains may be necessary. Automated water level controllers reduce time inputs, but are more costly. Raising the water table four feet in a sandy loam soil with drains 60 feet apart could take 3 to 5 days. Times would be longer for silt loam and clay soils.

Design

Information usually needed for a properly designed subirrigation system includes soil properties, topography, water supply, power supply, existing drainage specifications, crops to be grown, time available for system operation and management, and other information. The designer will determine the layout of the system, the depth and spacing of the drains, the pumping plant capacity, and the size and location of water control structures. The slope, hydraulic gradeline, and the size of drains must be determined for both subsurface drainage and subirrigation. An important final part of the design process is the economic analysis.

Problems of Subirrigation

The system operator or manager should be well aware of several potential problem that may occur with subirrigation. Sudden heavy rains during the irrigation mode may flood the crop root zone, especially if the weir setting in the water control structure is high (and thus a high water table in the field). When there is a high water table, there will be less water storage available in the soil. This problem may be solved by careful on-site management. The operator or manager should review weather patterns, and if possible allow time for the soil to partially drain before a rain occurs. This will help create some storage capacity in the soil for the expected rain.

Another major problem may be creating and maintaining a level water table throughout the field. This is especially true in soils with low lateral hydraulic conductivity, such as clays. Problems also exist in soils that lack an adequate restrictive layer below the drain depth. Careful site evaluation is very important before proceeding with design and construction.

Monitoring of Farm Water Tables

Water table monitoring can improve irrigation and drainage management for agricultural production in the high water table areas. In water table management, a target water table level is selected and water is either added to or removed from the field or farm according to whether the existing water table is lower or higher than the target, respectively. The water table depth and the amount it is allowed to deviate from the target level are dependent on soil properties, crop characteristics, pumping capacities, and the level of protection from flood or drought conditions desired by the grower. It is evident that water table depth is one of the most important physical features of a cropped field, and that its measurement is vital for optimum management.

Current public interest is directed towards achieving optimum water management for water supply and quality enhancement. Agricultural demands on water supplies and the negative effects of agricultural drainage water on the environment could decrease appreciably through better control of target water levels, permissible fluctuations, and pumping rates and durations. To achieve this, however, it is necessary to accurately track water tables both during and between pumping activities. Water table monitoring is vital to any management or research program that addresses farm water supply and quality issues in shallow water-table areas.

Water table monitoring, and its effective incorporation into a farm water management program, requires that four distinct activities be carried out. First, observation wells must be constructed and installed at strategic locations in the area to be monitored. Second, water levels in the wells must be accurately read and recorded on a regular basis. Third, the resulting data must be analysed and put into a readily useful form for the front line water managers. Finally, the resulting information must be incorporated into scheduling and managing irrigation and drainage events. Adherence to this program should enable a grower to optimise his water usage with water table management irrigation and drainage systems.

Presently, irrigation and drainage decisions are often made based on rainfall amounts and drive-by scanning of water levels in ditches (ditch-riding). Pumping decisions are then made according to past experience. This method seems inexact, but can be quite satisfactory, depending on the experience of the water manager. However using this method, there are as many ways to manage water tables as there are front line managers.

A water management policy maker can convey his expectations to his ditch riders, but what transpires during interpretation and implementation of those expectations may be unique to each ditch rider. To obtain some consistency and uniformity in making pumping decisions, water table levels should be measured and used in the decision making process. An additional benefit in the form of enhanced, documented knowledge of the farm water system and of localised and system-wide responses to pumping can be obtained through the use of simple, inexpensive water table level sensing instruments and a notebook.

Manual monitoring and recording methods generate useful information upon which to base irrigation and drainage decisions. The data can be used directly when a decision is needed. Direct or immediate use of the measurement can be made without the recording of values. A manager simply checks the water table level prior to making the decision to irrigate or drain.

Alternatively, the manager can implement a data collection and analysis program which would enable him to calibrate pumping amounts with ditch water levels and field water table behavior under various rainfall, irrigation, evapotranspiration, and drainage conditions. This program requires additional monitoring before and after the event, as well as regular monitoring between events, for optimum utility. Whatever level of monitoring sophistication is desired, manual water table measurements can remove much of the present guesswork involved in making pumping decisions.

Measurement tools for manual water table monitoring are inexpensive and can be constructed easily out of off-the-shelf components. Data management techniques are simple and effective, as long as the limitations of the methods are understood when the technique is selected.

It must be emphasised that manual monitoring methods have less utility than continuous recordings of water table levels. In-depth use of manual methods also requires more time commitment than present decision making processes. In addition, continual care by a conscientious operator is needed when taking measurements.

The water manager must determine the objectives of the monitoring program prior to system selection. A commitment of time and resources to data management and interpretation must be made while considering system alternatives. For instance, if monitoring the water table level is only a substitute for estimating ditch water levels, the simplest monitoring program would be adequate. However, if relationships between pumping, rainfall, irrigation, drainage, ditch water levels, and water tables are desired, more sophisticated monitoring programs will be required.

Observation Wells and Initialisation

After the well is constructed, initialisation, or establishment of a datum or reference measuring point on the well casing, is required. The vertical distance from the measuring point to the ground surface must be determined. For convenience, the initialisation procedures applicable to manual measurements are summarised here.

When a water table measurement is taken , it must be reported as a depth above or below a selected point of interest. For field and farm monitoring, the ideal point of interest is the ground surface. However, the ground surface is extremely variable, and changes throughout the cropping season. A single point on the ground surface may not be representative of the average field elevation. In fact, a measurement from a point on the ground surface cannot be considered to be anything more than the depth to the water table at that point. Therefore, it is necessary to establish a permanent datum, or reference point, from which all water table depths will be measured during the monitoring period. Obviously, the datum will then have to be related to the ground surface.

For convenience, the reference point is often chosen to be the top of the observation well casing. The tops of the well casings frequently are not parallel to the ground surface. Hence, it is advisable to mark the exact location using an indelible marker or by notching the top edge of the casing. Any future manual measurements should be read relative to the same location by holding the measuring device against the mark or notch. It is essential that the well casing be secured firmly in place so that the measurement reference point will not move during the monitoring period.

Relating the datum to the ground surface is more complicated. This can be done by simply determining the distance of the marked point above ground at the base of the well casing. However, this may not be entirely correct, especially since the soil around the casing was disturbed during well installation. The proper method of datum establishment is to survey the area around the observation well. If several wells around the farm will be monitored and data compared, their elevations should be measured with respect to a common benchmark.

Several ground elevations should be measured in a grid pattern around each observation well. Care must be taken to place the surveying rod lightly on the ground in order to avoid erroneous readings due to the rod penetrating the ground surface. The readings should be taken without any attempt to smooth the ground surface before placement of the surveying rod.

For bedded crops, the ground surface elevation should be taken at the top of the bed. Additional rod readings may be desired in the furrows between beds to track the consistency of bed height. For sugarcane and other non-bedded crops, the ground surface should be the average of the ridges and furrows formed during field planting and disking operations. Small grid spacings will ensure that both ridges and furrows will be included in the readings. It is unnecessary that the rod readings be true elevations with respect to sea level. Any arbitrary value can be assigned to the common benchmark. Ten feet or 10 meters is often used as the benchmark elevation for this type of differential leveling.

The elevations at each point on the grid are measured and then averaged around each well location. The resulting value is the average ground surface elevation for that well. If the same procedure is used around all the other wells in the field or on the farm, the average ground elevations at each site can be related to each other.

While measuring ground surface elevations relative to the benchmark, a sighting should also be taken on the measurement point marked on the top of the well casing. The elevation of the measurement point relative to the benchmark, minus the elevation of the average ground surface relative to the same benchmark, yields the height of the measurement point above the average ground surface.

The lack of surveying equipment may preclude such measurements. If such is the case, the ground around the observation well can be smoothed and a surface representative of the average ground surface can be estimated. When using this procedure, it should be remembered that the water table depths measured at any two sites cannot be compared. The water table readings will represent the water table depth only at the point measured. An assumption is then made that there is a certain area over which that ground elevation and water table depth is applicable. The limitations of this method are obvious. However, if the purpose of the water table measurement is to obtain a gross estimate of the water table in the general area, and if the resultant measurement will be used only in immediate decision making, the technique is acceptable.

Measurement Instruments

The methods of water table monitoring discussed herein require that a person visit the well site and manually measure the water table level. The instruments described are tools that can be used to actually sense the position of the water table. The means of recording data and data sheet formats are left to the discretion of the user.

There are many ways to manually measure the depth of a shallow water table. Virtually any item that is rigid, such as yardsticks or folding rulers, can be lowered into an observation well until it makes contact with the water surface in the well. Depth from the datum can be read directly if the tool is graduated. Alternatively, the tool can be marked, removed from the well, and laid against a measuring instrument to determine depth. As an alternative to the rigid tool, a flexible string or tape can be used in the same manner as long as it does not deform greatly when stretched to eliminate kinks.

The manual instruments discussed are tools that ease the process of water table measurement or sensing. It is not difficult to dip yardsticks, folding rulers, or tape measures into observation wells. Although such activities can suffice, the procedure becomes tedious and time-consuming, discouraging rigid adherence to a monitoring program.

Some of the instruments suggested were designed to reduce the strain of determining when a stick or tape makes contact with the water surface in the observation well. Others were developed to reduce the time and effort needed for obtaining water table depths when quick estimates are needed for immediate decision making. All methods were developed with the intention of increasing the accuracy of judging water table depths in fields and around farms.

Measuring Sticks

The simplest instruments used for water table depth measurements are measuring sticks. This category includes ungraduated laths of wood, lengths of rigid PVC pipe, or metal rods with or without graduations marked on them, yardsticks, folding carpenter's rules, surveying rods, etc. All require an individual visit the well site on a regular basis, or whenever readings are desired. The measuring sticks are simply lowered into the shallow observation well until the lower end just touches the water surface. The device is then read adjacent to the measuring point marked on the top of the well casing. The depth of the measuring point above ground surface, determined during initialisation, is then subtracted from the reading. The resulting value is the depth of the water table below ground surface. The process is illustrated in Figure 4.

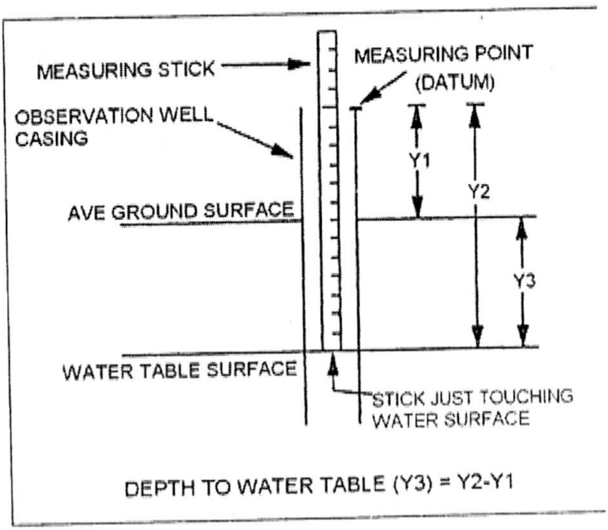

Figure 4. Calculation of the depth to water table from the ground surface.

When using measuring sticks, it is important to ensure that they are as nearly vertical as possible so the readings will not be distorted. A carpenter's rule, for example, tends to bow when extended. Additionally, the water table must be shallow enough, and the well diameter large enough, so that the operator can see the stick making contact with the water.

Obviously, this method is time consuming, labor intensive, and greatly depends on operator care for accuracy. However, the method is still more desirable than estimating ditch water levels and assuming the field water table level to be identical because ditch water levels and field water table levels do not generally coincide.

Another related method makes use of a tape measure or other length of flexible material. Standard metal tape measures can be used but are subject to many of the same limitations of the rigid sticks. It is recommended that a nylon or Fiberglass tape be used. These types of tape measures can be purchased in open reel configurations which will be useful during transport between well sites. An ordinary piece of string is not recommended because of its tendency to stretch and kink.

At the end of the tape measure, a small lead weight, or series of weights, should be attached to the metal anchoring piece found on most of these devices. Lead sinkers available in any fishing supply store are ideal. The weights will ensure that the tape hangs vertically. The tape is then lowered into the well until contact with water is made. Water table depth calculations are the same as for the above methods.

The above methods of measuring water table depths are the most basic ones. Obviously, they are better than using no monitoring device at all. The methods are all inexpensive, with the cost of a tape measure being the most significant expense. A single measuring device can be used for all wells. The methods, however, are tedious and time-consuming, and accuracy depends greatly on the diligence of the operator. The process of discerning instrument contact with the water level in the observation well can be difficult and can cause operator strain when large numbers of wells need to be read.

Well-sounding Devices

A simple well-sounding device can be used to ease the strain of determining instrument-water surface contact and to improve accuracy of measurements when using measuring sticks or tapes. The sounder is a set of electrical lead wires that indicate the water table by a visible light or audible tone when the electrical circuit is completed by submersion of the lead wires. Graduations on the lead wires allow accurate measurement of the depth to the water table. These tools are available commercially.

However, they are generally used for deep well applications such as determining depths of water in pumped wells. Hence graduations may not be as closely spaced as desired and the accuracy may be insufficient for shallow water table monitoring. Additionally, the user must buy the units capable of measuring 20 or more times the depth required for shallow water table applications. The unit consists of a paired set of graduated wires, a reel, and electronic circuits for driving the water contact indicator.

A simple well sounder can be assembled with off-the-shelf parts for less than $30. The electronic circuitry consists of lengths of light gauge wire (22 or higher), a 9 volt battery, with snap-on leads and clamp, a transistor

(2N3906), a Single Pole-Single Throw (SPST) switch, a 100 ohm resistor, a perforated circuit board with approximate dimensions of 2 in x 3 in, a Light Emitting Diode (LED) or buzzer, and a small plastic project box available at any electronic hobby store. Figure 5 shows the circuit schematic and the correct orientation of the transistor.

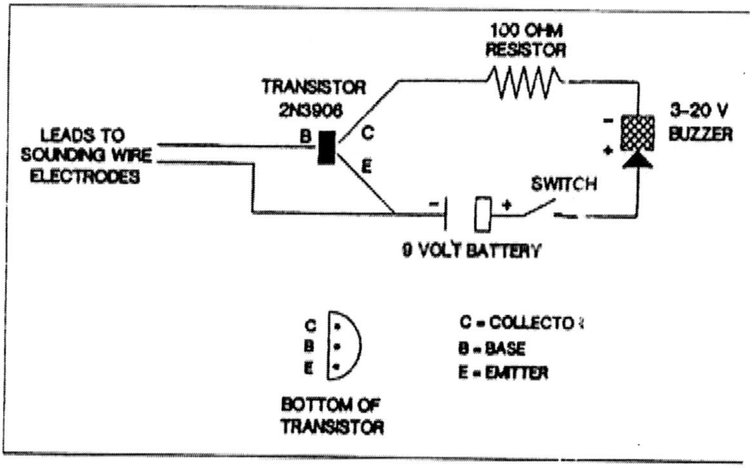

Figure 5. Well sounder circuit schematic

It is highly recommended that the LED be replaced by a buzzer. In brightly lit situations, such as those encountered in the field, an audible tone is much more appropriate than an LED. A PC Board Mount Piezo-Electric Buzzer with an operating range of 3 to 20 volts DC is an ideal replacement that requires few circuit adaptations. Also, the resistor should be changed to alter the tone and loudness of the buzzer. For maximum loudness without the sound becoming too shrill, a resistor in the 1000 to 2200 ohm range is effective. Alternatively 9 or 12 volt MiniBuzzers, also available at electronic hobby stores, can be used. The buzzers cost no more than $2.50

each. The circuit should be enclosed in the plastic box to protect it from the weather and abuse during handling. A folding carpenter's rule or other rigid graduated device can be used to lower the sensing device to the water table surface. The design shown in Figure 6 is recommended.

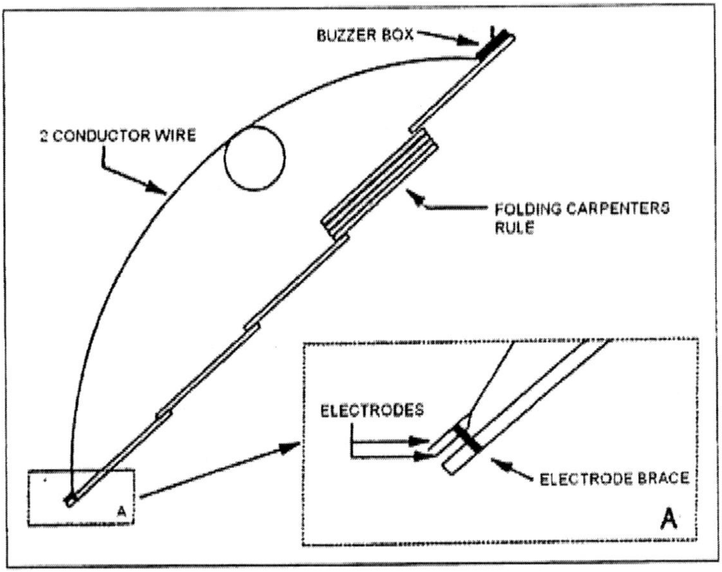

Figure 6. Well sounder constructed out of a carpenter's rule.

Two wires should be run the length of the stick. The ends of the wires must be stripped and separated so that they do not come into contact with each other or with a common wetted piece of the assembly during measurement. If such contact is made, the buzzer will sound continuously when the unit is switched on. For example, if the lead wires in Figure 6 were both touching the carpenters rule, water adhering to the stick may complete the circuit. The unit will then have to be dried out between readings.

The rigid stick and carpenter's rule can present problems during use. The carpenter's rule is flexible and may deform when lowered into the well. Additionally, when stored, it is folded up, causing kinking of the wire leads. The leads do not have to be attached to the rule, but tangles may occur when separate. The rigid yardstick or graduated PVC pipe configuration can be inconvenient when transporting the device through fields. It should be remembered that the total length of the device must be at least as long as the greatest depth from measuring point to water table surface expected.

The ideal configuration is to use a nylon or Fiberglass open reel tape measure. The circuit housing can be attached to the reel using screws. For best results, the plastic box should be attached to the winding crank, and the wires passed through a hole drilled in the crank and reel hub. In this configuration, the buzzer box and all wires will turn with the spool of tape thus avoiding kinking of the wires. These tape measures are available in 25 ft lengths and greater. The 25 ft length is more than adequate for most applications. The 50 ft tapes may be a simple matter to remove the tape and cut it to the desired length. In many cases, 10 ft will be adequate.

While the tape is disassembled for cutting, it should be stretched out on a flat surface. Lead weights should be attached to the metal anchoring assembly at the end of the tape measure. The lead wires from the electronic circuitry can then be tacked to the tape with epoxy or PVC cement as shown in Figure 7. The ends of the lead wires (the electrodes) should be set up as shown in Figure 7. When dry, the tape can be re-installed in the reel. It will probably be desirable to attach the lead wires to the tape prior to their connection to the buzzer unit. In this event, simply add connectors between the electrode wires and the buzzer lead wires. The connectors will facilitate servicing the unit since it can be readily broken down into two separate items.

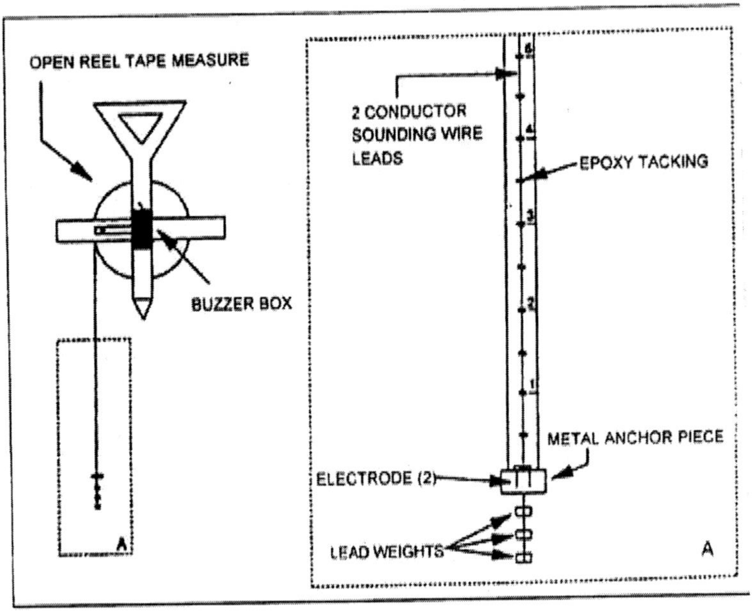

Figure 7. Well sounder constructed using an open reel tape measure.

The type of wire used with these reel tape measures is extremely important. The wire must be thin and flexible enough so that it will roll up on the reel and unwind without creating permanent kinks that prohibit the straightening of the tape. Weights added to the end of the measuring tape will help prevent kinking. Keep in mind that the longer the tape length that will actually be used, the more weight will be needed to ensure a straight tape. Multi-stranded telecommunications wire or ES-232 ribbon cable are two types of wire that work well for this application. The ribbon cable commonly comes in rolls of 25 ft with 25 conductors. Shorter lengths are equally common. When using this wire, two strands can easily be separated from the roll. It is not recommended that the two wires used be separated from each other.

Care must be taken so that the insulation around the wires remains intact. In the field, the weighted end of the tape is lowered into the observation well until the buzzer sounds. The tape is then carefully raised until the buzzer stops. Following that, the tape should be slowly lowered the until the buzzer sounds again. It is advisable that the tape be raised and lowered slightly once contact is made in order to determine the precise water table elevation. The point at which the buzzer first begins to sound indicates the water table location. The piezoelectric buzzer may continue to sound when not in contact with the water table. However, there is an audible difference in tone and intensity when the electrodes come in direct contact with water.

The tape measure should be read relative to the measuring datum point when contact is made. Performing the appropriate subtraction will again yield the depth of the water table below ground surface. This method of locating the water table considerably eases the strain of determining when a measuring device contacts the water table surface. Measurement becomes a simple matter of lowering a tape measure into the well, listening for the tone, and reading the depth. The tape is easily retracted from the well and can be transported without difficulty. This method of manually reading water tables is especially recommended when accurate readings are desired. The device is also extremely useful in initialising continuous type recorders.

Float Poles

When the intent of monitoring water tables is solely to obtain an approximate reading for immediate decision making, a different method may be preferable. A water manager may simply want to know how deep the water table is, without expending much more energy, time, or resources than in his present ditch-riding activities. In most cases, the data will not be recorded and no relationships will be drawn between different locations or between ditch water levels, rainfall events, drainage and irrigation responses, and field water tables.

The method makes use of a float device placed in the observation well. The device moves up and down with the movement of the water table. Attached to the float is a rigid graduated scale which moves with the position of the water table. The graduations on the scale can be marked so that the device registers actual water table position. The scale should be marked following similar procedures to those used to initialise the other manual devices. The distance above ground of the measuring point on the well casing is measured. Next, the depth to water table from ground surface is determined using one of the previously presented methods.

The float device with scale attachment is lowered into the well. The point on the rod adjacent to the measuring point is then marked and the depth noted as the measured depth minus the height of the measuring point above ground. From the marked point on the rod, use a measuring tape or yardstick to mark graduations above and below the initial reading.

As the float rises, the depth to water table decreases. Hence, the scale numbers marked, should decrease as graduations continue down towards the float. A zero water table mark will be encountered. Marks below this point should be marked with a plus sign, or differently, to indicate pounding on the surface. As shown in Figure 8, the numbers below the zero point should increase.

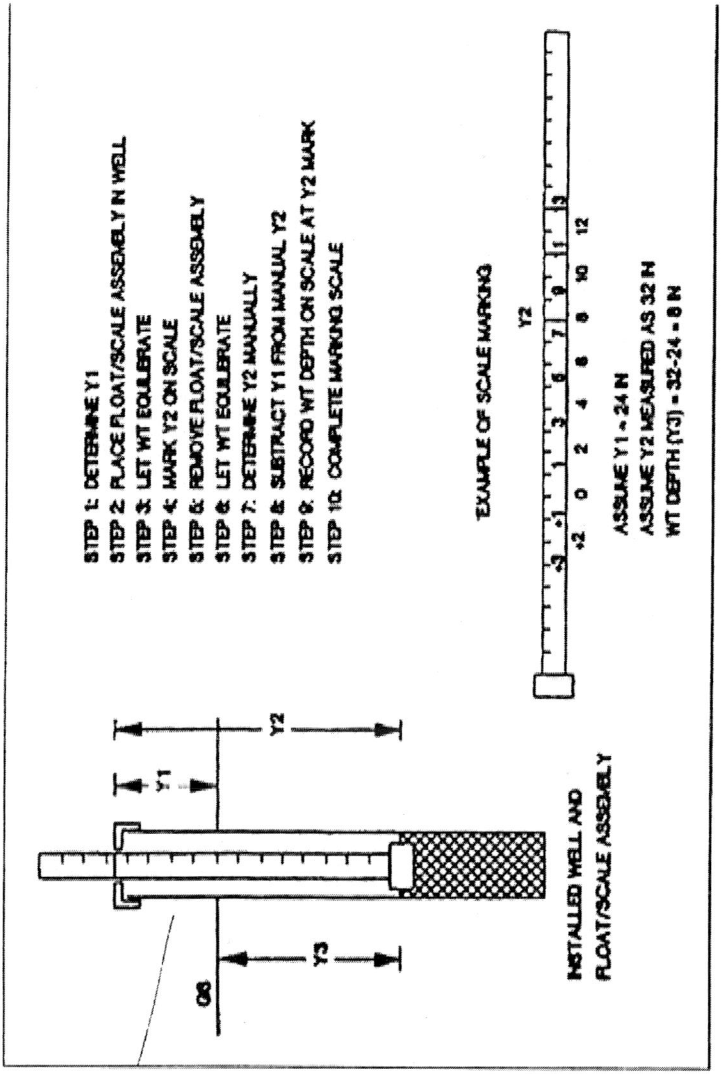

Figure 8. Marking graduations on float/scale or float pole.

Markings above the initial reading should continue with the scale increasing.

Simply measuring distances and marking the scale relative to the bottom of the float will result in incorrect measurements since the float will settle in the well until buoyant forces equal the weight of-the float-scale assembly, much as a boat settles in water.

The float assembly is then placed in the well. The scale must be supported at the top of the well in order to keep it vertical. A metal supporting brace can be fashioned out of a wire coat hanger as shown in Figure 9. Alteratively, a hole slightly larger than the scale rod can be cut in a PVC cap (Figure 9). The cap should then be placed over the rod scale and secured to the well casing. In this case, the distance of the top of the cap above the well casing measure point must be measured and added to the original measuring point to ground surface height. Alternatively, surveying readings should be taken on top of the cap when placed firmly over the well casing.

As an alternative to having a float and rod assembly, the entire unit ran be constructed out of PVC pipe as shown in Figure 10. A length of PVC pipe, smaller in outside diameter than the well casing inside diameter should be used. Both ends of the pipe should be capped. It is important that the bottom cap be fastened securely using PVC cement or another weatherproof adhesive.

It is recommended that the seal not be formed with silicone caulking since caulking will eventually deteriorate and separate from the PVC, resulting in possible water entry into the assembly. Sealing the upper cap is not as critical as long as it is clearly marked. Possible confusion could otherwise occur during marking of graduations, resulting in the lesser sealed end being placed in the well. The cap on top of the float pole is

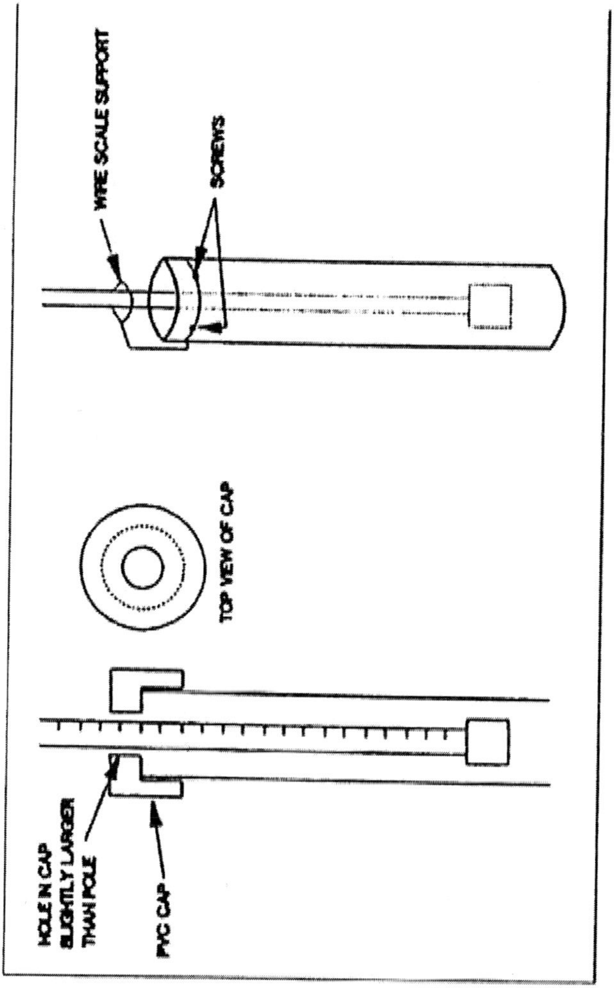

Figure 9. Two suggestions for constructing float/scale or float pole guides/supports.

necessary to prevent precipitation from filling the float. This assembly acts as the float and scale of the system.

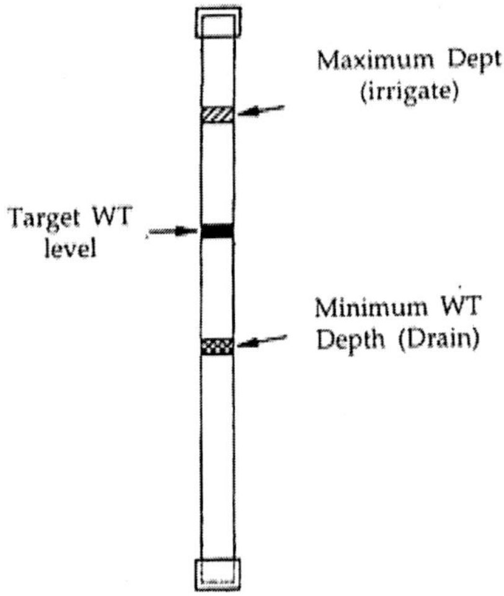

Figure 10. PVC float pole marked with target and critical water table levels.

A hole slightly larger than the outside diameter of the float pole should be drilled in the PVC cap on top of the well casing. The float pole should then be marked with graduations as was done for the previously discussed float scale assembly. The float pipe is then inserted into the well cap which is then secured to the casing. It is recommended that the well casing cap not be cemented to the casing so that removal of the unit is possible if maintenance of the well or float is necessary.

Several methods of enhancing the readability of the above mechanisms are available. As it is, the user must get within close proximity of the well to read the numbers. Color schemes can be used to facilitate readings from a distance. For example, alternate colors that will stand out against expected background colors can be used for every other 1 ft or 6 in increment. Sighting on the poles can be further enhanced by using a pair of binoculars. The critical water table elevations for signifying when an irrigation or drainage event should occur can be marked on the pole as shown in Figure 11.

Any other critical points of interest can also be marked accordingly. If this is done, the well/monitoring assembly can yield much information with little more equipment and work than is presently necessary. This system also enables the policy makers to set water table fluctuation levels that are easily communicated to, and carried out by, front line managers. This system will work well for short-height crops. When crops such as corn or sugar cane are involved, the method may not work or may be difficult to implement because of the height of the crop. Naturally, the reading point on the well casing cap must be higher than the maximum crop height or readings will be impossible.

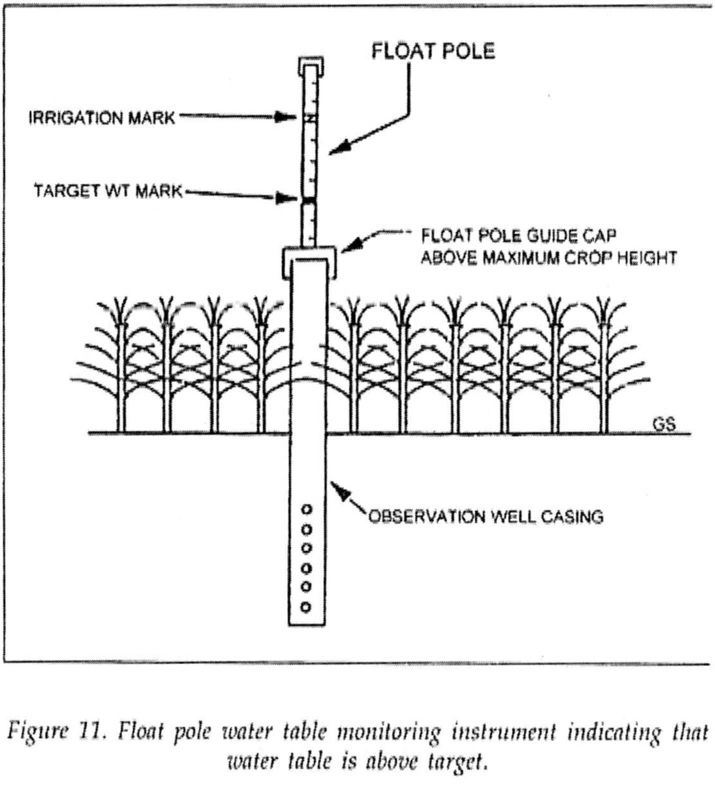

Figure 11. Float pole water table monitoring instrument indicating that water table is above target.

If crops are tall, problems may occur. For example, if the well casing needs to be 10 ft above ground surface, and the well depth is only 4 ft, the well casing may tend to tip over unless braced. The other problem occurs in initialisation. This problem, too, can be overcome by measuring the distance from the top of the casing cap to a marked point on the casing, whose elevation can be surveyed once the unit is installed. That fixed amount must also be accounted for in marking the scale.

Another problem that might occur, especially in the PVC float assembly, is restriction of the float travel. For example, if the observation well top is only 1 ft above ground, and water fluctuations range from flooded to 4 or 5 ft, it is likely that the float will get hung up at high water table levels. When a large length of PVC float is out of the well, the float will tend to lean. If the lean is not prevented by the well casing, the sides of the hole in the casing cap will impede the movement of the float. This problem can be remedied by increasing the height of the observation well above ground, using a larger diameter float pole, or increasing the size of the guide hole in the well casing cap.

The float pole method of water table monitoring can be a valuable tool for systems that continue to use ditch-riding techniques. Rather than estimating ditch levels and assuming they are indicative of field water levels, the ditch rider can now note critical water table elevations in the field centers. Additionally, the manager can terminate an irrigation or drainage event based on actual water tables in fields rather than having to wait until ditch levels and field water tables equilibrate to make a judgment. The user can also enhance his management of water tables by gaining a more quantitative feel for water table responses to pumping and rainfall.

Other Methods

Other methods of manually reading water tables exist. In fact, the only limiting factor is one's imagination. Some methods make use of electronic items such as potentiometers and pressure transducers. These can be mounted on a well and be read manually with a simple multi-meter whenever a reading is desired. Additionally, strip chart recorders can be placed in an observation well and read only when data is desired. It is assumed, though, that if these devices are to be installed, the user will have intentions for data uses above what is discussed here because of their costs and calibration and initialisation processes.

Monitoring schedules depend on the desires of the operator. The techniques discussed can offer valuable data without exceeding the present level of effort expended in ditch-riding activities. It may, however, be possible to keep tighter controls on water tables by taking daily readings. The readings should be taken at the same time each day for maximum accuracy since water tables tend to drop during the day and recover over night. If the user intends to apply the data to determining relationships between rainfall, drainage, irrigation, ditch water levels and field water tables, daily readings are virtually mandatory. However, recording data need not be done daily unless desired.

Measurements and recordings should occur more frequently before and after rainfall, or before, during, and after irrigation and drainage. The time interval between readings is left to the discretion of the grower, and will depend on his desire to draw meaningful relationships. Obviously, more measurements yield more data upon which to base relationships. Yet collecting more data is a waste of time and resources unless the user has a well-defined use for it. Recording measurements on multi-day or weekly intervals has little meaning unless constant water tables are being maintained by automated pump stations. In this case again, it will be beneficial to take more readings around pumping and rainfall events.

The observation wells should be checked regularly to ensure that they are functional. Common problems encountered are filling in of the well casing with soil, insect nests in the casing impeding float travel, and destruction of the well by farm machinery. The rigid measuring instruments should be checked for deformations. Electronic sounders are easily checked by dipping the electrodes in water. A multi-meter may be useful to determine which circuit component needs replacement if the unit is not operating. Float poles should be checked to ensure that the marked scales have not rubbed off. It is also necessary to check the float scale guide to ensure that it is supporting the float scale in a vertical position.

4

Drainage Water Assessment

Drainage water management refers to the management and control over the quantity and quality of the drainage water generated in an agricultural drainage basin in arid and semi-arid areas and its final safe disposal. This is achieved through irrigation water conservation measures and the reuse, disposal and treatment of drainage water. Managing drainage water at the field, irrigation-scheme and river-basin levels entails a number of activities including:

— regulating water table levels in the drainage system to ensure the maintenance of favourable soil moisture conditions for optimal crop growth and salinity control;

— developing irrigation and drainage water management strategies to ensure that disposal regulations and water quality standards are met and dealing with issues that might occur;

— setting distribution priorities and criteria for reuse in water scarce areas; and

— establishing cost sharing imposed on stakeholders for the use of poor quality water and required treatment to meet the water quality standards for drain discharge.

The main reasons for developing a drainage water management strategy are:

(i) prevention of economic and agricultural losses from waterlogging, salinization and water quality degradation;

(ii) concern for quality degradation of shared water resources; and

(iii) the need to conserve water for different water users under conditions of actual or projected water scarcity.

In addition, the need to comply with drainage water policies and regulations can provide a strong incentive for improved drainage water management.

Many countries and states, e.g. Australia, India, Egypt and California, the United States of America, have a drainage water disposal policy and drainage effluent disposal regulations. In California, the United States of America, and in Australia, the drainage policy guidelines consist of difficult but achievable targets with an active enforcement of regulations. India and Egypt have policy guidelines of a general nature but have not reached the maturity of Californian and Australian laws.

Law enforcement often fails in these countries, mainly as a result of administrative shortcomings and unrealistic quality guidelines for their conditions and resources. In these countries and in countries where clear laws and regulations are absent, the prevention of economic and agricultural losses and the conservation of water for other beneficial uses will normally be the main driving factors behind the development of a drainage water management strategy.

Conservation Measures

A major goal of conservation measures is to reduce the volume of drainage effluent generated and the mass discharge of salts and other constituents of concern while at the same time saving water for other beneficial uses. Conservation measures can directly affect the need for and extent of reuse as well as the quantity and quality of drainage effluent requiring disposal and/or treatment.

Where competition for water quantity and quality among different groups of users is a major issue and where drainage effluent disposal is constrained or threatens ecologically sensitive areas, conservation measures are among the first to be considered. The various conservation measures include: source reduction; shallow groundwater table management; groundwater management; and land retirement. Table 1 provides a short description of the four conservation measures and some points for consideration.

The first line of action in on-farm water conservation is source reduction or reducing deep percolation. Where the goal is not achievable by source reduction, it may be necessary to implement other conservation measures in combination with reuse measures. As different crops require different amounts of water for optimal crop growth and as rooting patterns differ among crops, source reduction and water table management could be combined with a change in cropping pattern to optimise the desired effect of the measures. Changes in cropping pattern should only be considered where compatible with the broader development objectives for the area under consideration.

Reuse Measures

The major aim of reuse measures is to reduce the amount

Table 1. Conservation measures, practices and points for consideration

Option	Practices	Points for consideration
Source reduction	Reduce the volume of deep percolation through: improving irrigation performance by surface irrigation, changing from surface to precision irrigation methods, modifying irrigation schedules, upgrading irrigation infrastructure, etc.	Farm- and system-level costs of system improvements have to be considered against the regional benefits. Salinisation and concentration of toxic elements in the rootzone need to be prevented by guaranteeing minimum leaching.
Shallow groundwater table management	Encourage the use of shallow groundwater to meet crop evaporation through maintaining high groundwater tables and practising deficit irrigation.	There is a danger of salinisation and concentration of toxic elements in the upper soil layers due to induced capillary rise. Moreover, the risk of insufficient aeration during rainfall in the rootzone needs to be addressed.
Groundwater management	Pumping from vertical wells could control water tables. Pumped water of ade-quate quality could serve as a substi-tute for surface water.	Prevent overexploitation of groundwater resources. There is a danger of intrusion and upwelling of saline groundwater. As groundwater often contains elevated concentrations of salts and trace elements, their buildup to toxic levels and soil degradation needs to be prevented.
Land retirement	Retire or fallow irrigated lands that are heavily affected by waterlogging and salinity or those lands that generate drainage effluent with extremely high concentrations of salts and/or other trace elements.	Retired lands can become excessively salinised and high concentration of toxic elements can build up in the topsoil preventing natural vegetation from establishing itself. Contaminated bare lands can affect productivity and be a health risk to humans in the vicinity due to wind erosion.

of drainage effluent while at the same time making additional water available for irrigation and other purposes. Reuse measures comprise:

— reuse in conventional agriculture;

— reuse in saline agriculture;

— Integrated Farm Drainage Management (IFDM) systems;

— reuse in wildlife habitats,

— wetlands and pastures; and

— reuse for initial reclamation of salt-affected lands.

Table 2 provides a brief description of the alternative reuse measures.

Reuse measures can be implemented in combination with conservation measures. Where the drainage water is of relatively good quality, its reuse potential in conventional agriculture is high. Where it is moderately to highly saline, its reuse may be limited to salt tolerant plants.

In California, the United States of America, a new IFDM system reuses drainage water sequentially in high water table lands with no opportunities for disposal of subsurface drainage. Freshwater is used to grow salt sensitive crops, and subsurface drainage water from it is reused to grow salt tolerant crops. Drainage water from the salt tolerant cropland is used to irrigate salt tolerant grasses and halophytes. When the drainage water is no longer usable, it is disposed into solar evaporators for salt harvest.

In this system and others, it is always necessary to generate a minimum volume of drainage effluent to prevent the rootzone from becoming too contaminated for any beneficial water use activity. Similar systems are being developed and tested in Australia.

Table 2. Reuse measures, practices and points for consideration

Option	Practices	Points for consideration
Reuse in conventional agriculture	Agricultural drainage water is collected and redistributed among farmers. Reuse can be direct or in conjunction with other sources of irrigation water. Conjunctive use can be through blending or by cyclic use of drainage water and other sources of irrigation water.	The extent of reuse depends on the quality of the drainage effluent, time of availability, crop tolerance, etc. Soil quality degradation and production losses need to be prevented through mitigation measures. Residuals of reused waters, which are often highly concentrated with a reduced volume, need to be managed.
Reuse in saline agriculture	Moderately and highly saline drainage water is collected and used to cultivate salt tolerant shrubs, trees and halophytes.	Sustainability of the saline agricultural systems needs to be safeguarded while the highly concentrated drainage effluent needs to be managed.
IFDM System	On-farm sequential reuse of agricultural drainage water on crops, trees and halophytes with increasing salt tolerance. In every reuse cycle the volume of drainage water decreases while the salt concentration increases. The final brine is disposed in a solar evaporator. Salt utilisation might be feasible.	To maintain adequate control of soil salinity and sodicity and to prevent a buildup of toxic elements, the leaching fraction must be sufficient. The solar evaporator should be designed in such a manner that it will not sustain aquatic life and attract birds. The salts have to be disposed of in a safe and sustainable manner.

Contd . . .

Contd . . .

Reuse in wildlife habitats & wetlands	Where of suitable quality, drainage water may be utilised to support wildlife, including waterbirds, fish, mammals and aquatic vegetation that serves as food and cover for wildlife.	Of primary concern is the possible presence of trace elements that may be toxic to wildlife through bioaccumulation in the food chain e.g. selenium, molybdenum and mercury.
Reuse for reclamation of salt-affected soils	Use of (moderately) saline drainage water for initial reclamation of saline, saline sodic or sodic soils. On sodic soils the saline water may help prevent soil dispersion and degradation of soil structure.	Once initial reclamation is obtained, water of sufficient quality and quantity needs to be available for desired land use. Initially, the drainage effluent will be highly saline and/or sodic and needs to be disposed of in a safe manner.

Treatment Measures

Drainage water treatment in a drainage water management plan normally takes place only under severe constraints, such as stringent regulations on disposal of saline drainage waters into streams, or severe water shortage. The drainage water treatment options are based on physical, chemical and/or biological processes. Many of these processes are borrowed from water treatment processes for drinking-water, sewage and industrial wastewater. A few are new processes to remove selenium. The water quality requirements of the treated water need to be well understood prior to the selection of any treatment measure.

Table 3 provides the array of drainage water treatment options.

Table 3. Array of drainage water treatment options

Description	Processes	Constituents removed or treated
Physical/ chemical	Sedimentation/ coagulation	Remove sediments and associated nutrients, pesticides and trace elements in sedimentation ponds with or without coagulants.
	Adsorption	Remove soluble constituents onto surfaces of adsorbents.
	Ion exchange	Exchange constituents with another using ion exchanger resins or columns.
	Reverse osmosis	Under pressure, separate out dissolved mineral salts through semi-permeable membranes.
	Coagulation/ precipitation	Use chemicals such as alum to coagulate or precipitate constituents of concern.
Biological	Reduction/ oxidation	Reduce oxidised mobile forms such as selenate to reduced immobile forms such as elemental selenium through microbially mediated processes.

Contd. . .

Contd. . .

Volatilisation	Some plants and microbes are capable of taking up constituents such as selenium and volatilising the methylated forms into the atmosphere.
Plant/algal uptake	Certain terrestrial plants and algae are capable of extracting large amounts of constituents such as selenium, nitrate and moly-bdenum.
Constructed flow-through wetlands	Constituents such as selenium and heavy metals are removed from drainage water. For selenium, the principal removal mechanism is reduction to elemental selenium and organic forms in the detrital matter. For heavy metals, the principal sink mechanism is sorption or fixation on the sediments.

The high costs of many of the treatment processes make them unsuitable for agricultural reuse. Treatment measures such as reverse osmosis are normally only considered for high-value purposes such as drinking-water supply. One option is to partially treat saline waters to a level that could be used for agriculture. However, the cost of brine management and disposal remains a major problem.

An exception to the typical high cost is the treatment of drainage water through constructed wetlands. Investigations are underway in California, the United States of America, in which wetland cells are planted with cattails, saltmarsh bulrush, tule, baltic rush, saltgrass, smooth cord grass and rabbitsfoot grass. The cells remove about 60-80 percent of the mass of selenium with hydraulic residence times ranging from 7 to 21 days. This treated water is then disposed into agricultural evaporation ponds with reduced toxic selenium effects on waterbirds. Constructed wetlands appear to have some potential to protect aquatic ecosystems and fisheries either downstream or in closed basins.

Disposal Measures

Even after the successful implementation of conservation and reuse measures, there will always be a residual volume of drainage effluent requiring disposal. Disposal options depend mainly on the situation of the drainage outlet in relation to natural disposal sites such as rivers, streams, lakes and oceans. Disposal options to surface water bodies comprise discharge into rivers and streams, river mouths and constructed outfall drains for direct disposal to oceans or saline lakes, and evaporation ponds.

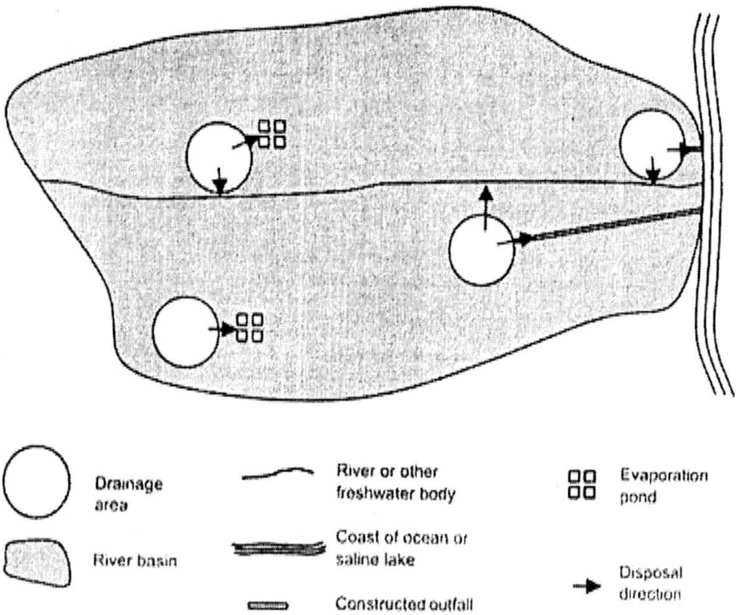

Figure 1. Options for disposal to surface water bodies

Table 4. Disposal measures, practices and points for consideration

Option	Practices	Points for consideration
River discharge	The drainage effluent is released into rivers, streams, etc. from where it finds its way to an ultimate salt sink, i.e. oceans and salt lakes. This option is especially appropriate during high discharge periods.	Disposal in rivers and lakes should not unduly impair other downstream uses including water required for sustaining fragile aquatic ecosystems. River disposal is often limited by disposal regulations.
Evaporation ponds	Disposal of drainage water in natural depressions or specially designed unlined basins. The impounded water dissipates through evaporation and inadvertent seepage losses, and deposits salts and trace elements.	Concentration of trace elements could adversely affect birds and wildlife through bioconcentration in the aquatic food chain to toxic levels. Adverse toxic impacts should be mitigated through special measures. Excessive seepage losses may pose a serious contamination risk to groundwater resources.
Outfall to saline lakes or oceans	Constructed main drain or disposal into river mouths to discharge effluents into the ocean or saline lakes.	Construction of outfall drains over long distances is normally an expensive undertaking and should only be considered where other alternatives are not feasible or the stream water quality is fragile. Disposal into oceans, bays and estuaries may be restricted if toxic trace elements are present. Disposal into tidal rivers needs tidal gates to prevent saltwater intrusion during high tide.
Deep-well injection	Treated drainage water is often injected into deep underlying permeable substratum in confined aquifers.	Formation of microbial slimes and colloidal particles may affect permeability in the stratum. There is a risk of seepage of poor quality water into fresh groundwater bodies.

Another potential disposal option is deep-well injection. However, in California, this approach failed due to the slow permeability of the geologic strata from the plugging of conducting pores. Besides the natural drainage conditions, the suitability of each of these measures depends on: the quality and quantity of drainage water requiring disposal; environmental and health risks; available technology and resources; and economic considerations. Table 5 provides a short description of the disposal measures.

Drainage water disposal into natural surface water bodies should entail minimal deleterious impacts on other downstream water uses and wildlife. Disposal on land in closed basins and agricultural evaporation basins should avoid undue harm to the ecology, particularly aquatic biota including fish and waterbirds.

Non-physical Drainage Water Management

In order to be fully effective, non-physical management measures, i.e. policy and legislation, should accompany physical drainage water management options. Different countries have implemented several policies, and these are often part of a general pollution policy. One of the most frequently mentioned principles with respect to pollution is that of the polluter pays. This means that economic policy instruments are applied to change the behaviour of farmers in such a way that pollution is minimised or that at least the polluter pays for its effects.

Weersink and Livernois provide an overview of such economic policy instruments for resolving water quality problems from agriculture. This overview was compiled for pollution by nutrients in humid temperate areas, such as Canada. However, some of these policies have relevance to pollution from the drainage of irrigated lands. The main problem of policy instruments in drainage water management is that water quality degradation from agriculture is non-point source pollution. Direct links between agricultural practices and water quality degradation are often difficult to determine and quantify.

Table 5. Economic policy instruments

Base	Target	Form
Performance of agricultural systems	Emission	Charges
	Ambient	Charges
		Salinity permits
Agricultural practices	Inputs	Charges
		Subsidies
	Outputs	Charges
		Subsidies

Weersink and Livernois divided the policy instruments into two groups:

(i) those based on the performance of an agricultural system, i.e. on the actual amount of pollution that caused; and

(ii) those based on agricultural practices such as fertilizer and agrochemical uses.

In the latter case, the underlying assumption is that certain agricultural practices will lead to more or less pollution. These policy instruments try to influence the practices and, indirectly, the pollution resulting from these practices.

Emission Levels

Fees are levied on the discharge of the polluter into the water. The aim is to stimulate the polluter to adopt practices that minimise pollution or to make the polluter pay for some of the damage caused. This requires measurement of individual discharges, which is often costly and not always practical to implement. Measurements at the individual farm level are probably only feasible with modern and large farms, such as in the United States of America and Spain. However, the case of the Panoche Fan shows that it remains difficult to establish equitable fines. Discharges from upstream farms with natural internal drainage cannot be measured while they may contribute significantly to the pollution in downstream regions. An alternative to measuring and levying fees on individual farms is to implement it at district level or at that of a water users association. This has the disadvantage that farmers applying conservation measures may still pay for the pollution caused by badly managed farms in the same district.

When basing fines on emission levels, there should be a decision as to whether to base them on concentrations of polluting elements or on the total load. Concentration is the water quality parameter used in drinking-water standards or for the health of aquatic animals and plants. Placing only concentration limits on discharge might encourage dilution or inefficient water use. Where both concentration and load limits are enforced, they tend to promote efficient water use.

Ambient Levels

Basing charges on the pollution in the downstream receiving water body can sometimes be a feasible alternative to basing them on the emission level, i.e. where the source of pollution is extremely difficult to measure. For example, in the case of pollution of groundwater, individual contributions to the pollution are difficult to measure and thus difficult to quantify. A disadvantage of using ambient levels under these conditions is that all farmers, including those applying sound water management practices, are charged equally.

Another case where this policy measure may be an attractive option is where the carrying capacity of the receiving water body changes continuously. For example, in India during the monsoon, river discharge is high and effluent disposals during this period increase downstream concentrations only slightly. On the other hand, during the dry season when river flow is low, emissions affect water quality substantially. In this case, using the ambient criterion reflects the major concern: maintaining river water quality for downstream users. This method stimulates corrective measures during critical periods.

Salinity Permits

An interesting solution for salinity management on a regional level is that adopted in the Murray-Darling basin in Australia. States within the basin have to meet electrical conductivity (EC) levels at the end of their river valleys, this in order to maintain a favourable water quality in the entire downstream river.

In order to reach this goal, a system of salt credits and debits is used. Credits are obtained for the implementation of any works that reduce the salinity in shared rivers. Debits are incurred based on the estimated shortfall in protecting shared rivers. The balance of credits and debits is registered for each state, and as a general principle each state must be in credit. The credits and debits are converted to EC impact at a location in the downstream area of the basin. This method allows states and catchment management authorities to decide on the most cost-effective options for their area whilst contributing to the overall basin-wide river salinity management plan.

Charges on Inputs

Instead of charging for pollution outflow from irrigation systems, economic incentives can be directed at reducing the inflow, which will normally lead to a reduced pollution outflow. Thus, as a way of reducing pollution, input charges seek to conserve water and minimise the use of agrochemicals and fertilizers. Volumetric water pricing is the main way of providing an economic incentive to save water. However, research from Iran has highlighted some limitations of this approach.

In the Iranian case, to serve as a proper economic incentive to change to pressurised irrigation, the total costs of water would need to equate to 60 percent of the revenues of a wheat crop, while at the moment they stand at 5 percent. Such an increase is probably politically impossible and would reduce farm incomes drastically. Surface irrigation improvement is less costly than changing to pressurised irrigation. Therefore, volumetric water pricing may be an economic incentive to improve surface irrigation water management.

Charges on inputs should also consider the way in which farmers can reduce water consumption. The main changes farmers can make are: grow less-water-demanding crops; apply water saving technologies; and change over to rainfed farming (generally not practised in arid and semi-arid areas). All these responses lead to less water consumption, but sometimes also to significant drops in farm income level.

On the basis of computer simulations, Varela-Ortega et al. analysed the responses of farmers in Spain to increasing water prices. Results showed that in modern irrigation districts, where irrigation efficiency is already quite high, no changes in water demand would occur unless the prices became restrictively high and farmers needed to change to rainfed farming. In contrast, in areas with the older irrigation systems, farmers would have ample scope for improving their water use efficiency.

An alternative to volumetric pricing is tiered water pricing. In this system, farmers pay a lower volumetric price for a reasonably necessary amount of water to grow a certain crop. A higher volumetric price is charged for additional amounts of water. Such a system is explicitly designed to reduce the amounts of water that generate most of the drainage water. The reasonably necessary amount should include the minimal leaching requirement in order to maintain a favourable salt balance in the rootzone unless sufficient rainfall occurs in the area. A main problem would be to define the 'reasonable necessary amount of water'.

In Broadview Water District, California, the United States of America, tiering levels differ only between crops and not between soils, because farmers consider the latter approach to be inequitable. Tiering levels were established at 90 percent of the average depths applied in the years before the system was implemented. This system of tiered water pricing led to a decrease in water applied to five of the seven main crops. More importantly, the drained volume in 20 of the 25 subsurface systems in the water district decreased by 23 percent and the salt load declined by 25 percent.

Volumetric pricing policies are typically only applicable in on-demand systems. Other systems deliver fixed amounts of water to farms, and they fall outside the responsibility of the individual farmers. In such systems, more gains could be obtained by matching supply and demand more closely to each other in a technical/ operational way.

Subsidies on Practices

A changeover to water saving or pollution prevention measures might not always be the objective of individual farmers nor might it always be to their advantage. In the case of Iran, it was calculated that the investments in sprinkler irrigation are of the same order of magnitude as the value of the water that can be saved, if this is used to irrigate a larger area or high-value crops. However, if the water saved is diverted to irrigate land in other areas, upstream farmers will be investing while benefits accrue to other areas. Another example is the case of farmers located in the higher areas with free drainage in the Panoche Fan. Farmers in these areas contribute much to the pollution, but there is no direct incentive for them to invest in water saving or pollution prevention measures. In both cases, subsidies might stimulate farmers to invest in water saving technologies or pollution prevention measures.

Charging/subsidising Outputs

A final instrument proposed by Weersink and Livernois is a charge on the crops. In the case of nitrogen leaching, the charge focuses on crops that need large amounts of nitrogen, and thus have a higher risk of causing pollution. In irrigated agriculture, one could think of policies that discourage growing crops with large water requirements by charging the crops or by promoting deep-rooting crops or less-water-demanding crops through subsidies. A common example is where farmers pay a price per area depending on the crops grown.

Combined Measures

A single economic or policy instrument will probably not resolve water pollution problems originating from irrigated agriculture. Combinations of policies can help to enforce their effect. Especially in situations where farmers are inequitably charged, a combination of measures might help to charge farmers more equitably. For example, emissions from an irrigation district could be measured and fines applied where limits are surpassed. These fines could then be distributed amongst the farmers, applying a ranking based on how farmers have used their inputs or what measures they implement. The most inefficient users should then pay comparatively more than the most efficient ones.

Evaluation of Drainage Water Management Options

Normally, more than one drainage water management option will need to be considered and implemented to attain the desired objectives. In this case, interactions and trade-offs occur. Because of the complexity of the processes and interactions, computer models are best suited to selecting an optimal combination of drainage measures. The selection of a set of drainage water management measures requires more considerations than solely technical feasibility. Figure 2 shows a flowchart for selecting a set of drainage water management measures.

Step 1 is to gather technical, environmental, social, economic and institutional information about the site and the available drainage water management options. Step 2 is to evaluate the desirability of the options from a technical and environmental perspective. Typically, this step requires the use of computer models.

Step 3 involves ranking the site's options, based on the technical and environmental desirability. Step 4 involves an analysis of the relative economic efficiency of the various options. Marginal cost curves may be useful in determining which options are more efficient than

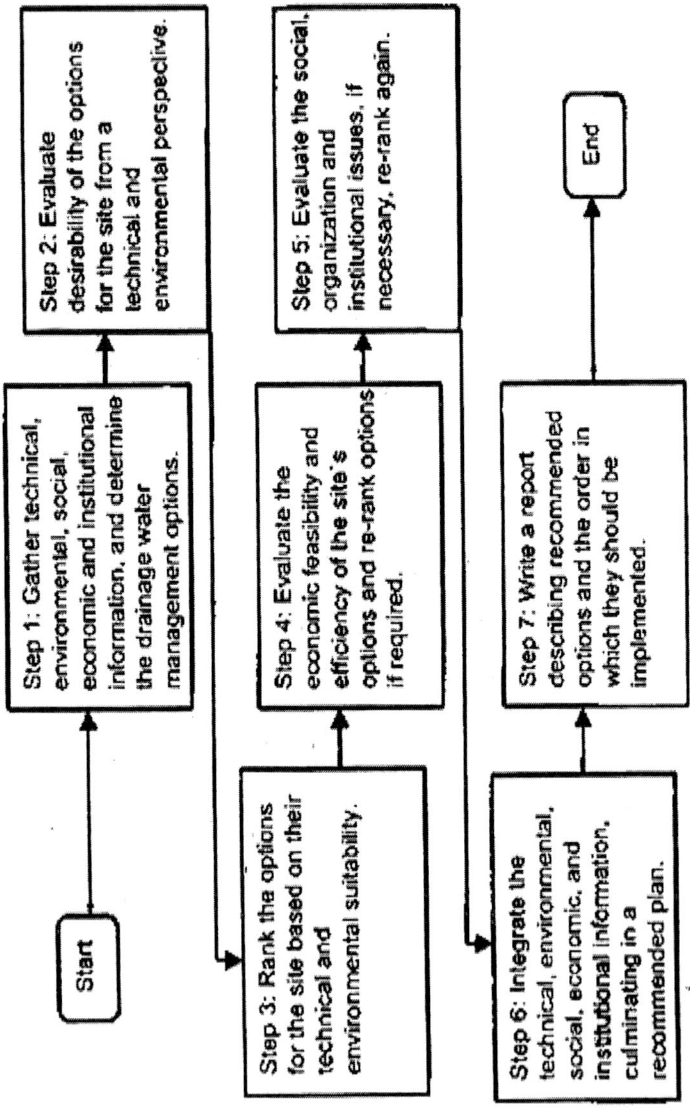

Figure 2. Flowchart of process for selecting an optimal set of drainage water management options

others. When evaluating the economic efficiency of a highly complex, multi-parameter system, it may be advantageous to turn to a computerised economic optimisation model. The course of action ultimately decided upon also depends on the social, organisational and institutional suitability of the proposed options and may run counter to what is economically efficient.

Step 5 evaluates the social and institutional issues. Step 6 is perhaps the most difficult as it always involves intuitive judgement and a certain measure of creativity. Collating all the information and developing a stepwise path for implementing the favoured options takes time and considerable thought.

Benchmarking

The Australian National Committee on Irrigation and Drainage is exploring another approach called benchmarking as part of its efforts to improve the performance of irrigated agriculture. The principal goal of benchmarking is to find and implement best practice for the organisation in question by learning about their own organisation through comparison with their historical performance and with practices and outcomes of others.

In 1998/1999, this committee evaluated 47 parameters in 46 water providers to gauge system operation, environmental issues, business processes and financial performance. New parameters to be considered with specific regard to on-farm water use include: security of water supply, water savings from operational and seepage remediation, metering of supplies, water trading, salt balance, water quality monitoring, and operating costs per unit length of channel or pipeline.

The benchmarking reports and compiled data have already led the water industry to consider and/or adopt improved technologies to enhance economic, environmental and social performance. This approach has received positive responses from the water industry, and the plan is to now evaluate additional comparative parameters, such as groundwater quality, absolute changes in water table, extent of water reuse and recycling, average water use on major crops, and financial reports on economic value added.

Quality Concerns in Drainage Water Management

There are several factors to consider when determining the opportunities for and constraints on the safe use, treatment and disposal of agricultural drainage water. Information and data desired at the site of drainage water production include: rate of drainage water production per unit area, chemical concentration of constituents of concern, and the rates of mass emission.

Drainage water management requires additional information and data on drainage water quality and its suitability for the intended water uses as well as an understanding of environmental and health concerns. Upstream drainage water management affects the needs and water quality requirements of downstream water users.

Table 6 provides a summary of changes in water quality expected in irrigation return flow relative to irrigation water applied. The expected differences in quality in the return flow are described relative to the supply water because actual concentrations in supply waters vary. The operational spill waters (bypass water) from distribution conveyances are not expected to differ much from the quality of the supply water except for some pickup or deposition of sediments. In contrast, surface runoff or irrigation tailwater tends to pick up considerable amounts of sediments and associated nutrients, phosphorus in particular, as well as water-applied agricultural chemicals such as pesticides and nitrogen fertilizers.

Table 6. Expected quality characteristics of irrigation return flow as related to applied irrigation waters

Quality parameters	Operational spills	Irrigation tailwater	Subsurface drainage
General quality	0	+	++
Salinity	0	0, +	++
Nitrogen	0	0, +, ++	++, +
Oxygen demanding organics	0	+, 0	0, -, -
Sediments	+, -	++	-
Pesticide residues	0	++	0, -, +
Phosphorus	0, +	++	0, -, +

0 not expected to be much different than the supply water.

+, - slight increase/pickup or decrease/deposition expected.

++ expected to be significantly higher due to concentrating effects, application of agricultural chemicals, erosional losses, pickup of natural geochemical sources, etc

- expected to be significantly lower due to filtration, fixation, microbiological degradation, etc.

Tailwater is typically similar to the applied waters in salinity and oxygen demanding organics, termed biochemical oxygen demand (BOD). Subsurface drainage is enriched in soluble components such as dissolved mineral salts and nitrates, very low in sediments, whereas other quality parameters are similar to the irrigation water. These changes in water quality of irrigation return flow depend on a number of factors including irrigation application methods, soil properties and conditions, application of agricultural chemicals, hydrogeology, drainage system, climate, and farmers' water management.

In many regions of the world, municipalities and industries discharge wastewater into open drains initially intended for the conveyance of only agricultural drainage and storm water. In developing countries especially, municipal and industrial wastewater is often insufficiently treated before disposal into such open drains.

The result is a risk that agricultural drainage water quality might be seriously contaminated with microbes, pathogens, toxic organics and trace elements including heavy metals. A knowledge of the composition of the drainage effluent and the ability to predict changes in the composition as a result of changes in crop, irrigation or drainage water management practices are important in the planning and management of drainage water.

Factors Affecting Drainage Water Quality

Geology and hydrology: The geology of the region plays an important role in drainage water quality. Through weathering processes, the types of rocks (both primary and sedimentary) in the upper and lower strata define the types and quantities of soluble constituents found in the irrigated area.

The oceans have submerged many parts of the continents during a period in their geological history. The uplift of these submerged geological formations and receding seas have left marine evaporites and sedimentary rocks behind, high in sea salts including sodium, chloride, magnesium, sulphate and boron.

These geological formations exist in varying thicknesses, depths and extents on the continents. Through hydrological processes, solutes can enter the upper stratum by irrigation or floodwater, upward groundwater flow in seepage zones, with rising groundwater levels, or capillary rise. Once the solutes are in the upper strata, they influence the quality of agricultural drainage water through farmers' irrigation and drainage water management. The following example shows how the geology and hydrology of an area influence the quality of agricultural drainage water. It also illustrates the relationship between geomorphology, waterlogging and salinisation.

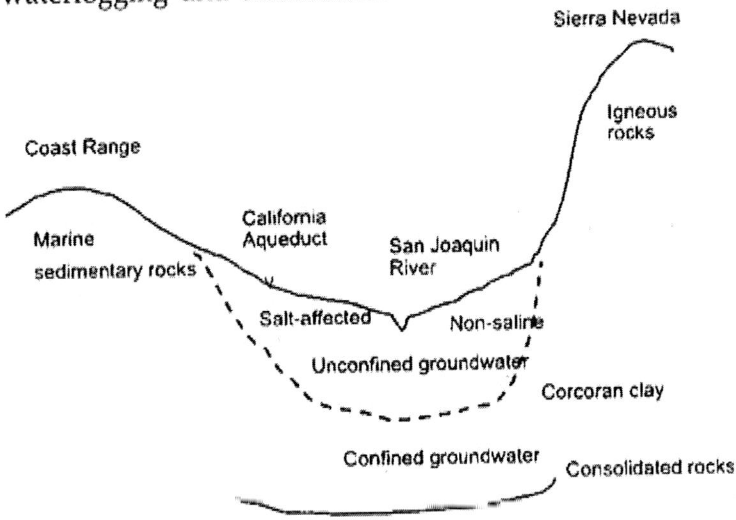

Figure 3. Cross-section of the San Joaquin Valley

Figure 3 shows a schematic cross-section of the San Joaquin Valley with the San Joaquin River as the principal drainage course for this river basin. The eastern side of the valley was formed from the alluvium of the Sierra Nevada, which consists mainly of granitic rocks. The soils derived from Sierran alluvium tend to be coarse textured and non-saline. The eastern groundwaters are characterised as low-salt calcium-bicarbonate-type water with total dissolved solids (TDS) typically in the 200-500 mg/litre range. In contrast, the soils on the western side were formed from alluvium of the Coast Range made up of uplifted marine sedimentary rocks. The soils on the western side tend to be finer textured and saline.

The groundwaters on the western side are characterised as moderately saline sodium-sulphate-type waters with TDS typically in the 1 000-10 000 mg/litre range. The unconfined aquifer in both sides of the valley is gradually being filled up with decades of irrigation deep percolation. The soils in the valley and lowest part of the alluvial fans in the western side are waterlogged and salt affected.

A nearly water-impermeable clay layer known as the Corcoran clay, about 200 m deep, serves as the boundary between the unconfined and confined aquifer. The groundwaters in the confined aquifer contain from 500 to 1 000 mg/litre TDS. During the geologic past, plate tectonics caused the horizontal-lying Corcoran clay in the shallow sea to tilt upwards forming the Coast Range.

Figure 4 is a freebody diagram of the waterlogged irrigated lands on the western side showing the water flow pathways in the surface and subsurface. The applied irrigation water is about 450 mg/litre calcium-bicarbonate-type imported water from the Sacramento River basin to the north. Much of the surface runoff is captured and reused on site.

Much of the collected saline subsurface drainage water (4 000-10 000 mg/litre TDS, sodium-sulphate type) is discharged into the San Joaquin River. Especially high concentrations of trace elements such as boron and selenium originating from the marine sedimentary rocks, found in the subsurface drainage waters, have given rise to environmental and health concerns. Discharges from these areas are now constrained by waste discharge requirements.

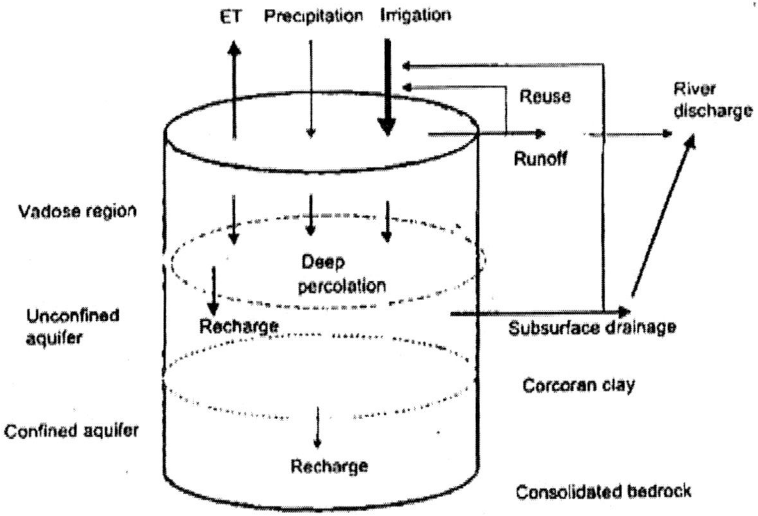

Figure 4. Freebody diagram of water flows in the San Joaquin Valley

A second example comes from the Aral Sea Basin, where in total 137 million tonnes of salt are annually discharged, of which 81 million tonnes (59 percent) originate from the irrigation water and 56 million tonnes (41 percent) from the mobilisation of salts from the subsoil. In the mid-stream areas, mobilisation of subsoil salts is the most substantial. The annual discharge from the Karshi oblast (Uzbekistan) is 10.8 million tonnes of salts, which corresponds to about 34 tonnes per hectare. About 4.3 million tonnes (about 40 percent) originate from irrigation water and the remainder are mobilised salts from the subsoil through irrigation and drainage. Further, in Australia, large amounts of salts are added to the soil profile by atmospheric deposition of salts from upwind wind erosion of salt pans.

Soils: Figure 5 depicts the water movement over the soil surface and through the soil profile. Soils serve not only as a medium for plant growth but also store water and nutrients and serve as the porous transport media. The soil's eroding capacity and chemical weathering leads to the generation of water-borne suspended particles and solutes, ranging from nutrients to all kinds of contaminants. Therefore, to understand the role of soils on drainage water quality, it is necessary to understand water movement over and through the soil and the associated suspended and dissolved substances it carries.

Of the water added to the soil, either in the form of rainfall or irrigation, part is lost through runoff and direct evaporation at the soil surface. Runoff water collects in natural and constructed surface drains from where it finds its way to the final disposal site (a river, evaporation pond or outfall drain to the ocean or saline lake). The other part infiltrates into the soil. This water fills up the soil pores and restores the soil moisture content up to field capacity under free drainage. The stored water is now available for plant root extraction to satisfy the water requirement of the crop.

Any water in excess of field capacity percolates below the rootzone to greater depth in the vadose zone. The deep percolation water may eventually serve as recharge to the groundwater or saturated zone. In irrigated areas with shallow groundwater tables, the recharge is immediate and causes the water table to rise. Where subsurface drainage is installed in waterlogged soils, the drainage system removes deep percolation and groundwater. Where the soil moisture content in the rootzone drops as a result of evapotranspiration and if there is no recharge from irrigation or rainfall, capillary rise into the rootzone might occur, depending on the water table depth, soil texture and structure, and seepage.

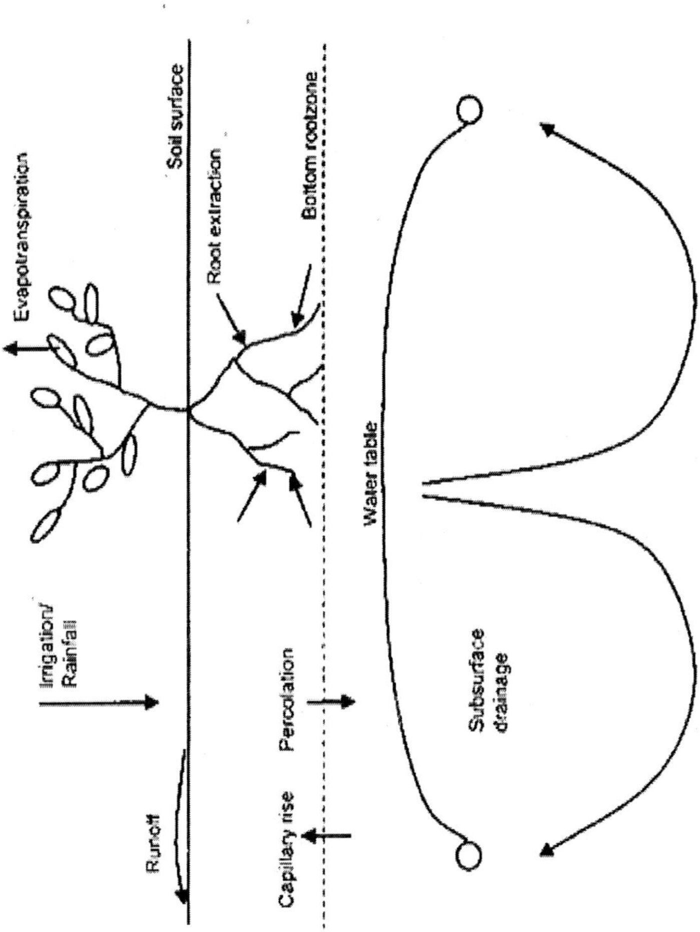

Figure 5. Water flow over and through the soil

Runoff in irrigated agriculture is mainly related to the intensity of irrigation and rainfall events in comparison to the infiltration capacity of the soil. Where the infiltration rate is smaller than the irrigation or rainfall intensity, water will accumulate on the soil surface and run off under a minimum surface slope.

Soil degradation, in terms of compaction and crust formation as well as cultivation on steep slopes, promotes surface runoff. Through the physical forces of the running water, soil particles become suspended in the water and are transported to open drains, ditches, streams, rivers and lakes.

Deposition of suspended sediments may occur downstream when current velocities decrease. Suspended soil particles are harmful to aquatic life as they diminish light transmission, but also because chemical contaminants may be associated with suspended sediments. Degradation of drainage water quality as affected by runoff from agricultural land is especially important in hilly areas and in areas with excess rainfall.

In the more arid areas and in flat plains, water flowing through the soil profile and associated solute fluxes affected by mineral solubility and adsorption processes are of more importance for the final quality of the drainage effluent than surface runoff. In some places, weathering of soil particles might play a major role in the quality of drainage water. However, in general the soil's ability to adsorb and release through ion exchange and transform chemical elements through microbially-mediated redox reactions plays a more important role.

In this context, soil characteristics such as water holding capacity, hydraulic conductivity, clay and organic matter content, soil minerals and soil microbes are important characteristics. Soil degradation in the form of erosion, compaction and loss of biological activity reduces the water and solute holding capacity of the soils. This increases the mobility of solutes through the soil and increases the risk that pollutants such as salts, nutrients and pesticides will be lost both to groundwater and through interception by subsurface drainage to surface water.

Climate: As the major transport of solutes through the soil is by the movement of water, climate plays a major role in determining drainage water quality. In humid tropics and temperate regions, the dominant movement of water through the soil is vertically downwards. Solutes, which are brought onto the soil by farmers or are naturally present in the upper soil layers, are leached into deeper soil layers and groundwater. Conversely, in arid climates where evaporation largely exceeds precipitation the dominant water movement through the soil is vertically upwards except during rainfall or irrigation events. Therefore, the chemical composition of deeper soil layers influences the quality of the shallow groundwater and the composition of the soil moisture in the rootzone. Climate and temperature also play a role in the rate of weathering and chemical processes.

Cropping patterns: Cropping patterns play an important role in the quality of drainage water in a number of respects. First, crops extract water from the rootzone resulting in an evapoconcentration of salts and other solutes in the soil solution. Where the solubility product of minerals is exceeded through evapoconcentration, minerals precipitate out. This changes the composition of the soil solution and thus influences the chemical quality of subsurface drainage waters. Second, crop residues add organic matter to the soil profile. Organic matter in the soil increases the adsorptive capacity for metals and other solutes. Furthermore, organic matter enhances the soil structure, which increases the water holding capacity of the soil. The organic matter also serves as a carbon source for soil microbes involved in transformations such as denitrification, sulphate reduction and methane production in submerged soils. Third, plants extract nutrients through their rooting system and some plants have the capacity to accumulate large amounts of certain salts and toxic elements. Fourth, as not all crops have the same salt tolerance the type of crop largely determines the maximum salt concentration in the rootzone and the amount of water needed to maintain a favourable salt balance in the rootzone. Last, incorporation of nitrogen fixing crops such as legumes can help to reduce nitrogen leaching. Legumes in symbiosis with nitrogen fixing bacteria are both users and producers of nitrogen. They can substitute chemical nitrogen fertilizer in the crop rotation. Deep-rooted perennial crops such as alfalfa can also help to prevent nitrogen leaching by absorbing large amounts of nitrogen.

Use of agricultural inputs: Application of fertilizers, pesticides, soil and water amendments, and animal manures may influence the quality of drainage water to a great extent. The amounts and timing of application in relation to the growing stage of the crops, timing of irrigation, drainage practices and applied soil conservation measures largely define the influence of fertilizer, amendment and pesticide application on drainage water quality. Furthermore, the characteristics of the fertilizers themselves play a major role, also on the possible contamination of drainage water.

Most nitrogen fertilizers are highly soluble and mobile in the soil, and nitrates readily enter drainage water through leaching processes. A portion of the nitrogen fertilizers made up of ammonium or anhydrous ammonia is initially adsorbed to the soil exchange complex but ammonium ions oxidise readily to nitrate. This is also true of urea containing nitrogen fertilizers that eventually oxidise to nitrates. Conversely, phosphorus fertilizers are less mobile in the soil because they have very low solubility, and phosphates are adsorbed on positive sites in soil organic matter and clay minerals.

The main route for phosphorous to drainage water is through runoff as sediment-bound inorganic and organic phosphate. Runoff waters may contain residues of anhydrous ammonia injected into irrigation water. Ammonia is highly toxic to fish. Runoff waters may also contain sediment-bound organic nitrogen. Excessive levels of nitrogen and phosphate in discharge waters may result in the eutrophication of water bodies.

Water and soil amendments such as gypsum may contribute significantly to salinity. Although gypsum is sparingly soluble, calcium contributed from gypsum may exchange with adsorbed sodium, and sodium sulphate is a highly soluble mineral. Acidic amendments react with soil calcium carbonates but do not contribute to salinity because the solubility of calcium carbonate is very low. Where animal manures are mineralised, nitrates and salts are produced. Some animal manure contains appreciable amounts of salts.

Irrigation and drainage management: Irrigation and drainage management are the main factors influencing the flow of water over and through the soil in arid zone croplands. As solute transport takes place mainly through soil water fluxes, irrigation and drainage management determine to a great extent the solute fluxes through the soil. In irrigation management, timing in relation to crop water requirements and to fertilizer and pesticide application is key to controlling the amount of soluble elements that will leach below the rootzone, from where they can be intercepted by subsurface drains. The timing of irrigation also affects capillary rise into the rootzone, which might cause an accumulation of salts in the rootzone but at the same time reduces the need for irrigation water. Of equal importance to the timing is the amount of irrigation water applied. Excess water, including infiltrated rainfall, leaches to deeper soil layers.

Drainage techniques and design: The choice of drainage technology and certain design choices influence the quality drainage effluent considerably. Subsurface drainage enhances the flow of water through the soil. Studies from the United States of America have shown that, in comparison to soils drained by means of surface drains, subsurface drainage reduces the amount of runoff and subsequently reduces the phosphorous contamination of surface water. At the same time, subsurface drainage enhances nitrogen leaching. Upon interception of leaching water, the nitrogen concentration of the final drainage effluent increases. In the same way, subsurface drainage intercepts other soluble elements present in the soil. This offers the possibility to control the concentration of harmful salts and toxic trace elements in the rootzone for optimal crop production but reduces the possibility for disposal and reuse of drainage water.

For water table and salinity control in irrigated lands, two types of drainage technologies are generally employed: subsurface tile drainage, and tubewell drainage. Subsurface tile drains are generally installed at a depth of less than 2 m below the soil surface. However, exceptionally, drains are installed at depths of up to 3 m.

Tubewell drains vary between 6 and 10 m but might reach a depth of 100 m, e.g. deep tubewell drains in Pakistan. These differences in drain depth influence the quality of the effluent. The quality of drainage water from subsurface tile drains is influenced by the quality of irrigation water, the applied farm inputs and the quality of the shallow groundwater.

In contrast, the quality of drainage effluent from tubewell drains is related mainly to the quality of the groundwater and to a lesser extent to the quality of irrigation water. Based on computer simulations in Pakistan, a comparison of tubewell and tile drainage showed that for Kairpur the salinity of drainage effluent was 11 750 and 2 100 ppm for tubewell and tile drainage, respectively, whereas for Panjnad Abbiana it was 22 918 and 610 ppm, respectively.

Table 7. Salt applied in irrigation water and removed by drains

Treatment	Salt removed(kg/ha)	Salt removed/Saltapplied*
Deep drains	607	11
Shallow drains	32	0.7

* Mean salt applied = 50 kg/ha

The depth of tile drain installation also influences the amount and quality of the generated drainage water. In the Aral Sea Basin, deep drains increase the flow that has to be evacuated and its salt content through mobilisation of fossil groundwater through deep flow paths. For example, drainage water in the Karshi oblast collector is extremely saline (up to 8 g/litre). Expected mineralisation levels of drainage water resulting from irrigation only are much less.

Experiments in Australia with shallow and deep drains in a vineyard on a clay soil led to the same conclusion. Less drainage effluent was generated with shallow drains (0.7 m deep) than with deep drains (1.8 m deep). Moreover, the shallow drainage water had a lower salinity than the deep drainage water; together with reduced drainage volumes this resulted in a reduced salt load. As the deep drains removed much more salt than imported with the irrigation water, this led to the conclusion that the salinity in the effluent of the deep drains was derived from deeper soil layers.

Model simulations by Fio and Deverel confirm these observations. They showed that the base flow towards tile drains increased and the quality decreased with increasing installation depth under non-irrigated conditions. They also showed that this is a result of the path flows becoming deeper and longer with increased drain depth and spacing. On the other hand, during irrigation events the discharge of the shallow and deeper drains became similar as recharge due to deep percolation increased and the proportional contribution of deep groundwater to drain lateral flow decreased.

Nature of Drainage Water Quality

Ions

The major cations and anions making up salinity are sodium, calcium, magnesium, potassium, bicarbonate, sulphate, chloride and nitrate. A lumped salinity parameter is frequently used such as EC in decisiemens per metre (dS/m) or TDS in milligrams per litre. The water quality of surface runoff typically deviates little from the composition of the irrigation water even if it flows over soils with visible salt crusts.

On the other hand, deep percolation displaces salts accumulated in the soil profile from natural chemical weathering, blown in by salt dust, as well as evapoconcentrated salts derived from the applied irrigation water. Thus, the salt content of the collected subsurface drainage water mainly reflects the salinity characteristics of the soil solution, which in turn is influenced by soil parent material, salinity of the shallow groundwater and salts brought into the soil with irrigation water. In many places, the drainage water composition is further influenced by the mineral composition of deep groundwater which is intercepted by the drains.

Toxic Trace Elements

Trace elements are commonly present at low levels in nature. Many trace elements are essential micronutrients in very small quantities such as iron, manganese, molybdenum and zinc, but the range between deficiency and toxicity is narrow. Trace elements of concern in drainage water from irrigated lands include: arsenic, boron, cadmium, chromium, copper, lead, mercury, molybdenum, nickel, selenium, strontium, uranium, vanadium and zinc.

The heavy metal trace elements are fixed strongly by soil materials and tend to be mobile only in the topmost soil layers. However, some of them form mobile metal-organic complexes in the presence of organic matter. Some trace elements (arsenic, selenium, molybdenum and uranium) are relatively immobile in the reduced form (precipitated or elemental) or are adsorbed while the oxidised and oxyanion species are mobile. For example, selenium is soluble in alkaline and well-oxidised soils.

Similar to the dissolved mineral salts, trace elements are evapoconcentrated in the presence of a growing crop as more or less pure water is lost into the atmosphere and the trace elements remain in the soil solution. Elevated concentrations of selenium, boron and molybdenum may be found in soils formed from Cretaceous shale and marine sedimentary rocks and their shallow groundwaters such as in the western side of the San Joaquin Valley. Selenium typically exists in a reduced form in geologic formations as seleniferous pyrite or organic forms of selenium. When the formations are exposed to the atmosphere, the reduced selenium oxidises into soluble forms and may be subject to transport from irrigation and precipitation.

Naturally occurring arsenic is commonly found in volcanic glass in volcanic rocks of rhyolitic to intermediate composition. It may be adsorbed to and coprecipitated with metal oxides, adsorbed to clay minerals and associated with sulphide minerals and organic carbon. In natural aquifers, arsenic is especially a problem in West Bengal, India, and Bangladesh but it reportedly also occurs in the United States of America, Hungary, Chile, China, Argentina, Ghana, Mexico, the Philippines, New Zealand and Mongolia.

To evaluate whether trace elements might potentially cause a problem in drainage water management, Westcot studied ranges and geometric mean values in soils for the priority pollutant trace elements. Comparing actual soil data with these levels would give an initial idea of the potential for trace element leaching.

Agropollutants

The two main agropollutants in agricultural drainage waters are nutrients and pesticides. Once a pesticide enters the soil, its fate is largely dependent on sorption and persistence. Sorption is mainly related to the organic carbon content of soils while persistence is evaluated in terms of half-life (the time taken for 50 percent of the chemical to be degraded or transformed). Pesticides with a low sorption coefficient and high water solubility are likely to be leached while pesticides with a long half-life could be persistent.

Pesticides vary widely in their behaviour. Pesticides that dissolve readily in water have a tendency to be leached into the groundwater and to be lost as surface runoff from irrigation and rainfall events. Pesticides with high vapour pressure are easily lost into the atmosphere during application. Pesticides that are strongly sorbed to soil particles are not readily leached but may be bound to sediments discharged from croplands. Pesticides may be chemically degraded through such processes as hydrolysis and photochemical degradation. Pesticides may be biologically degraded or transformed by soil microbes.

Therefore, many pesticides are mainly found in surface drainage water and not in subsurface drainage water as such, due to the filtering action of the soils. However, some pesticides have a tendency to be leached through soil profiles and accumulate in groundwater, such as organophosphates.

Sediments

Sediment contamination is a main concern for surface drainage in hilly areas and in areas with high rainfall. Sediment production in arid zones occurs in improperly designed and managed surface irrigation systems, especially furrow irrigation. Sediments are a direct threat to living aquatic resources and the aquatic environment in general. They also increase the cost of drinking water treatment and maintenance of open surface drainage networks from sediment deposition like in Pakistan and

China. In addition, phosphate, organic nitrogen and pesticides bound to sediment particles are a source of pollution.

Sediment production can be reduced by minimum tillage practices and limiting surface runoff through sound irrigation practices. Sediment settling ponds may be used to reduce the load of sediments in receiving waters. Polyacrylamides appear to serve as an excellent coagulant for sediments in farm drainage. Sediments are not normally found in subsurface drainage water. However, a drainage pipe filled with soil particles might cause sediment pollution in subsurface drainage water.

Water Quality Concerns for Water Uses

Crop Production

The total concentration of salts in drainage effluent is of major concern for irrigated agriculture. Salinity in the rootzone increases the osmotic pressure in the soil solution. This causes plants to exert more energy to take up soil water to meet their evapotranspiration requirement. At a certain salt concentration, plant roots will not be able to generate enough forces to extract water from the soil profile. Water stress will occur, resulting in yield reduction. The extent to which the plants are able to tolerate salinity in the soil moisture differs between crop species and varieties.

For the stability of the soil structure, the composition of the soil solution is an important factor. In the solid phase, soils have a net negative surface charge. The magnitude of the cation exchange capacity (CEC) depends on the amount and type of clay and the organic matter content.

Cations such as calcium, magnesium, sodium, potassium and hydrogen are adsorbed on the exchanger sites. Normally, a large fraction of the adsorbed cations is divalent calcium and magnesium. Divalent cations adsorbed to clay minerals provide structure and stability. Where monovalent cations dominate the exchangeable cations (sodium in particular), the soil structure looses its stability and structural degradation occurs easily.

As cations are mutually replaceable, the composition of the exchangeable cations is related to the proportion of cations present in the soil solution. Therefore, where drainage water reuse for irrigation purposes is under consideration, not only the total salt concentration should be taken into account, but also the sodium to calcium and magnesium ratio, commonly expressed as the sodium adsorption ratio (SAR). High bicarbonate waters tend to precipitate out calcium carbonate. This may increase the SAR in the soil solution and increase the exchangeable sodium percentage (ESP) on the CEC.

The composition of the salts is also important for crop growth. Dominance of certain ions might cause an imbalance in ion uptake. This results in deficiencies of certain elements and depressed yields. The presence of high concentrations of sodium inhibits the uptake of calcium, causing nutritional disorders. Other ions can be toxic, causing characteristic injury symptoms as the ions accumulate in the plant. Toxic elements of major concern are chloride, sodium and boron.

The extent to which crops suffer from salinity stress depends on several factors. Although yield reductions are defined as a function of the average salt concentration in the rootzone, interactions between soil, water and climatic conditions influence the relationship. Exceedingly high air temperatures may cause a reduced salt tolerance. Cultural practices also determine to a certain extent yield reduction resulting from salinity stress. Other plant characteristics determine their ability to cope with salinity stress.

The variations between crops in salt tolerance are attributable to the fact that certain crops can make the necessary osmotic adjustment to enable them to extract more water from saline soils. This adjustment involves two mechanisms: absorption of salts from the soil solution, and synthesis of organic solutes. Halophytes tend to absorb salts and impound them in the vacuoles, while organic solutes serve the function of osmotic adjustment in the cytoplasm. Normal plants tend to exclude sodium and chloride ions. For this reason, these plants need to rely more than halophytes on the synthesis of organic osmolytes. As a result, they are more salt sensitive than halophytes.

Sensitivity to salts changes considerably during plant development. Most crops are sensitive to salinity during emergence and early development. Once established, most plants become increasingly tolerant during later stages of growth. There is general agreement that the earlier the plants are stressed, the greater the reduction in vegetative growth.

Not all trace elements are toxic and small quantities of many are essential for plant growth. However, excessive quantities might accumulate in plant tissues and cause growth reductions. Crop tolerance to trace element concentrations varies widely. When accumulated in plant tissue, certain trace elements are also toxic to animals and humans upon eating, e.g. selenium, arsenic and cadmium. As plants do not absorb most of the trace elements that are present in the soil, the trace elements accumulate in the soils.

Aquatic Resources

Aquatic organisms have different requirements with respect to the chemical and physical characteristics of a water body. Dissolved oxygen, adequate nutrient levels, and the absence of toxic concentrations of hazardous elements are essential factors for sustaining aquatic life. Drainage water disposal can disturb the chemical and physical characteristics of the aquatic habitat.

In natural water, the levels of trace elements are normally very low. Elevated concentration levels have a negative impact and harmful effects on aquatic life. Some trace elements such as mercury and selenium are of particular concern because of their bioaccumulative nature, even at very low concentrations. For example, in the United States of America the regulatory maximum contaminant level for selenium for aquatic biota in freshwaters is 2 ppb and for drinking-waters for humans, 50 ppb. The former is lower due to the bioaccumulation of selenium through the aquatic food chain.

Pesticides may also cause toxicity problems in aquatic organisms in surface waters. While pesticide use is currently highest in North America and Europe, it is expected to increase at a faster rate in developing countries in the near future. Many of the synthetic organic compounds are persistent and bioaccumulate. They magnify up the food chain and are often absorbed in body fat, where they can persist for a long time. In the case of fish tissues and fishery products, some of these compounds may also reach consumers. As fish are an important source of protein, it is essential to prevent and avoid accumulation of contaminants in fish or shellfish.

All aquatic organisms including fish or other aquatic resources living in contaminated water bodies are being exposed daily to a multitude of synthetic chemical compounds that disrupt the development of the reproductive, immune, nervous and endocrine systems by mimicking hormones, blocking the action of hormones, or by other unknown interference with the endocrine system. Fish have different life stages: egg, larvae, fingerling and adult. Various pollutants may have different effects on their life cycle and on their functions and abilities.

The greatest threat to the sustainability of inland fishery resources is degradation of the environment. According to the GEO-I prepared by the UNEP, access to and pollution of freshwater are among the four key priority areas. Various guidelines have been proposed for water important for fisheries or protecting aquatic environmental quality in general. Water quality guidelines established for temperate regions should not be applied without caution to other climate conditions as toxicity, persistence and accumulation rates might differ substantially.

Livestock Production

Water for livestock watering should be of high quality to prevent livestock diseases, salt imbalance, or poisoning by toxic constituents. Many of the water quality variables for livestock are the same as for human drinking-water resources although the total permissible levels of total suspended solids and salinity may be higher.

Concerns for Human Health

The quality of water has a major influence on public health. Poor microbiological quality is likely to lead to outbreaks of infectious water-borne diseases and may cause serious epidemics. Chemical water quality is generally of lower importance. The impact of chemicals on human health tends to be of a chronic long-term nature, and there is time available to take remedial action. However, acute effects may be encountered where major pollution events occur or where levels of certain chemicals are high from natural sources.

Increases in salinity, related to drainage water disposal on a shared water resource, may threaten its use for domestic and drinking-water supply. Although the WHO has not formulated any guidelines based on TDS, high salt concentrations can cause taste problems. Concentrations of less than 1 000 mg/litre (1.56 dS/m) are normally acceptable to consumers. For the majority of the major ions, no health guidelines have been derived. Present guidelines are based on taste and other side-effects of individual ions, such as staining of laundry by iron, or the rotten egg smell of sulphidic water.

Most of the toxic trace elements are included in the health criteria for guidelines for the quality of drinking-water as some of them are carcinogenic. For example, arsenic contamination of drinking-water supplies is of major concern in Bangladesh. Expected concentrations in natural waters are generally well below 1 mg/litre. Where concentrations are exceeded, expensive treatment processes are required to make the water acceptable for human consumption.

Some well-known chemical pollutants that affect health include nitrate, arsenic, mercury and fluoride. In addition, there is an increasing number of synthetic organic compounds released into the environment whose effect on human health is poorly understood, but appears to be carcinogenic. Humans also use water resources for bathing and recreation. Such activities in contaminated waters pose a health risk due to: the possibility of ingesting small quantities; contact with the eye, nose and ear; and contact through open wounds. Health risks related to recreation are mainly related to pathogenic contamination. The potential risks from chemical contamination of recreational waters are usually small. Even repeated exposure is unlikely to result in ill effects at the concentrations of contamination found in waters and with the exposure patterns of recreational users. However, the aesthetic quality of recreational water is extremely important for the psychological wellbeing of users.

5

Agricultural Water Reuse

In areas where irrigation water is scarce, the use of drainage water is an important strategy for supplementing water resources. The reuse may help alleviate drainage disposal problems by reducing the volume of drainage water involved. The reuse of drainage water for irrigation can reduce the overall problems of water pollution. Reuse measures consist of: reuse in conventional agriculture; reuse to grow salt tolerant crops; reuse in wildlife habitats and wetlands; and reuse for initial reclamation of salt-affected lands.

Drainage water is normally of inferior quality compared to the original irrigation water. Adequate attention needs to be paid to management measures to minimise long-term and short-term harmful effects on crop production, soil productivity and water quality at project or basin scale. The drainage water quality determines which crops can be irrigated. Highly saline drainage water cannot be used to irrigate salt sensitive crops, but it can be used on salt tolerant crops, trees, bushes and fodder crops. A major concern in reuse measures is that drainage water from reused waters is often highly concentrated, requiring careful management.

Drainage water quality is the major concern in reuse possibilities as it defines which crops can be irrigated and whether long-term degradation of soil productivity is a major issue. On the other hand, the soil type, drainage conditions of the land and the crop salt tolerance define what quality drainage water can be used for irrigation in combination with the availability of other freshwater resources.

Pollutants from surface runoff, i.e. sediments, pesticides and nutrients, play a minor role in reuse for crop production. However, for sustainable agricultural practices and to prevent environmental degradation, nutrients supplied with reused drainage water should be deducted from the fertilizer requirements in order to prevent imbalanced and excessive fertilizer application.

Subsurface drainage water generally shows increased concentrations of salts and sometimes certain trace elements and soluble nutrients. Salts and trace elements play a major role in the reuse of drainage water. Above a certain threshold value, high total concentrations of salts are harmful to crop growth, while individual salts can disturb nutrient uptake or be toxic to plants. A high sodium to calcium plus magnesium concentration ratio may cause unstable soil structure.

Soils with unstable structure are subject to crusting and compaction, degrading soil conditions for optimal crop growth. Toxic trace elements such as boron can interfere with optimal crop growth and others such as selenium and arsenic can enter the food chain when crops are irrigated with water containing high concentrations of these trace elements. This is of major concern for human and animal health.

Municipalities and industries often use agricultural main drains to dispose of their wastewater. In many areas around the world, municipal and industrial wastewater is either insufficiently treated, or not treated at all. Bacteriological and organic and inorganic compounds seriously pollute main drains, posing an environmental hazard to both human and wildlife and restricting reuse from main drains. For example, in the Nile Delta, Egypt, the discharge of untreated wastewater is a major concern for the sustainability of an irrigated agriculture that depends heavily on the large-scale reuse of agricultural drainage water to meet water shortfalls.

The biochemical oxygen demand (BOD) is defined as the amount of oxygen consumed by microbes in decomposing carbonaceous organic matter. The chemical oxygen demand (COD) is the amount of oxygen required to oxidize the organic matter and other reduced compounds. The high chemical versus biological oxygen demand (COD/BOD) ratios imply significant industrial pollution. Total coliform is the most probable number of faecal coliform (MPN) in 100 ml and a high value indicates severe pollution from municipal sewage water.

The degraded water quality threatens the expansion and even the continuation of the reuse of drainage water from the main drains (official reuse) in the Nile Delta. Since the 1990s, many reuse mixing stations have been under increasing pressure of water quality deterioration. Indeed, since 1992, 7 of the 23 main reuse mixing stations have been entirely or periodically closed.

Due to the increasing deterioration of water quality, new opportunities for reuse of drainage water are being explored. Between the centralised official reuse and the localised unofficial reuse (where farmers pump drainage water from the collector drains directly onto fields), there is the option to capture drainage water from branch drains and pump it into the branch canals at their intersections. This level of reuse, termed intermediate reuse, captures relatively good quality drainage water before it is discharged into the main drains where it is lost for irrigation because of pollution from other sectors. This problem highlights the need for catchment level planning to protect and use all water resources sustainably.

Maintaining a Favourable Salt Balance

The major concern in the reuse of agricultural drainage water is the buildup of salts and other trace elements in the rootzone to such an extent that it interferes with optimal crop growth and degradation of the aquifers. Applying more water than necessary during the growing season for evapotranspiration can leach the salts. In areas with insufficient natural drainage, leaching water will need to be removed through artificial drainage. As safe disposal of agricultural drainage water is often the hindrance to sustainable drainage water management, the solution in reuse lies in applying just enough water to maintain a favourable salt balance.

Maintaining a Favourable Soil Structure

The sodium hazards of irrigation water are related to the ability of excessive sodium or extremely low salinity concentrations to destabilise soil structure. The primary processes responsible for soil degradation are swelling and clay dispersion. Provided that the salt concentration in the soil water is below a critical flocculation concentration, clays will disperse spontaneously at high exchangeable sodium percentage (ESP) values, whereas at low ESP levels inputs of energy are required for dispersion.

The salt concentration in the soil water is crucial to determining soil physical behaviour because of its effects in promoting clay flocculation. However, the boundary (ESP/salt concentration of the soil water) between stable and unstable conditions varies from one soil to the next and changes with the clay mineralogy, pH, soil texture, and clay, organic matter and oxide content.

The short-term effects of irrigating with water having excess sodium or very low salt concentrations relate mainly to infiltration problems. The reduction of infiltration can be attributed to the dispersion and migration of clay minerals into soil pores, the swelling of expandable clays and crust formation. The potential infiltration problems in relation to the quality of the irrigation water are normally evaluated on the basis of the salinity and SAR of the irrigation water.

Various authors have developed stability lines related to the total salinity concentration and the SAR. The actual line that represents the division between stable and unstable soil conditions is unique for each soil type and varies with soil conditions. Published guidelines on infiltration problems in relation to the SAR and the salinity of the applied water can therefore provide only approximate guidance. Where large-scale reuse of drainage water is planned and sodium hazards might be expected, stability lines need to be established for local conditions.

Infiltration problems caused by the sodicity of irrigation water also depend on irrigation and soil management. Especially at lower SAR levels when chemical bonding is weakened, but no spontaneous dispersion takes place, inputs of energy are required for actual dispersion. For example, sprinkler irrigation increases the likelihood of surface crusting due to the high physical disruption as the drops hit the soil surface aggregates. In soil management, incorporation of organic matter increases the stability of the soil aggregates and reduces the hazards of structural degradation as a result of sodicity.

Reduction of the hydraulic conductivity in the soil profile is normally a long-term effect resulting from the use of sodic water. In particular, the presence of carbonates and bicarbonates in the water could result in soil degradation in the long term because precipitation of calcium carbonate increases soil SAR. Calcium in the form of calcite is one of the first salts to precipitate. Upon further concentration magnesium salts will also precipitate.

There is little need to undertake action to increase infiltration and or hydraulic conductivity unless crop water or leaching requirements cannot be met or if secondary problems reduce crop yields or impede seedling emergence. Secondary problems include crusting of seed beds, excessive weed growth and surface water ponding that can cause root rot, diseases, nutritional disorders, poor aeration and poor germination.

Management options to mitigate and reduce these problems can be chemical, biological and physical. Chemical management options entail adding chemical amendments to soil or water and thereby changing the soil or water chemistry. The aim of biological methods is to improve soil structure or to influence the soil chemistry through the decomposition of organic materials. Physical methods include cultural practices to increase infiltration rates during irrigation and rainfall or to prevent direct contact between ponding water and plant stems, roots and seeds.

Chemical Soil and Water Amendments

The aim of applying chemical amendments to soil or water is to improve poor infiltration caused by either a low salinity or by excessive sodium. The problem is most severe at low electrolyte concentration and high SAR. Improvements can be expected if the soluble calcium content is increased or a significant increase in salinity is achieved.

Most soil and water amendments supply calcium directly or indirectly through acid that reacts with soil calcium carbonates. Acid is not effective where calcium carbonate is not present in the soil profile. However, calcium carbonate is often present in arid soils. Water amendments are most effective when infiltration problems are caused by low to moderate saline water (EC < 0.5 dS/m) with a high SAR. Where moderate to high saline water (EC > 1.0 dS/m) causes problems, soil amendments are more effective.

Amendments need to react quickly if they are to solve actively infiltration problems caused by irrigation water. Many of the amendments for improving or reclaiming sodic soils react only after oxidation. Thus, they are slow and less suitable for solving infiltration problems caused by irrigation water. Slow reacting amendments include acid forming substances such as sulphur, pyrite, certain fertilizers and pressmud from sugar-cane factories.

Gypsum ($CaSO_4.2H_2O$) is the most commonly used amendment. It can either be applied to the soil or irrigation water. Infiltration problems normally occur primarily in the upper few centimetres of the soil. Hence, it is normally more effective to apply small frequent doses of gypsum left on the soil surface or mixed with the upper few centimetres of the topsoil than higher doses incorporated deeper in the soil profile.

The application of gypsum to irrigation water to prevent infiltration problems usually requires less gypsum per hectare than when it is used as soil application. Gypsum application to water is particularly effective when added to low salinity water (< 0.5 dS/m) and less effective for higher salinity water because of the difficulty of obtaining sufficient calcium in solution. In practice, it is not possible to obtain more than 1-4 meq/litre of dissolved calcium, or 0.86-3.44 g/litre of pure gypsum in fast-flowing irrigation streams. In low salinity water, these small amounts of dissolved calcium ions may increase the infiltration rate by as much as 300 percent. To use gypsum as a water amendment, finely ground gypsum (< 0.25 mm in diameter) is preferred as it has a higher solubility. This is also normally the purer grade of gypsum. A drawback is that finely ground purer grades of gypsum are more expensive, which often prevents small farmers from using it. The coarser grinds and lower grades of gypsum are more satisfactory for soil application.

Experiments have been conducted with the placement of gypsum rocks in the watercourse. The problem is that the amount of calcium dissolving from the gypsum rocks is low so the effectiveness depends on the stream velocity and volume. Experiments from Pakistan have shown that sufficient head and length of the supply canal from the tubewell to the watercourse are required to dissolve enough calcium in tubewell water. Where technically feasible, the use of gypsum stones has proved to be financially attractive for farmers. In large-scale applications, a drawback might be the cost of canal maintenance, as the gypsum blocks need to be removed before mechanical cleaning.

Sulphuric acid is an extensively used amendment for addressing infiltration problems. It is effective only where lime is present in the soil surface. This highly corrosive acid can be applied to the soil directly, where it reacts with lime making the naturally present calcium available for exchange with adsorbed sodium. Sulphuric acid reacts rapidly with soil lime making it a useful amendment to combat infiltration problems. When added to the irrigation water, it is neutralised by the carbonates and bicarbonates in the water and any excess would contribute towards dissolving the soil lime.

Biological Soil Amendments

Crop residues or other organic matter left in or added to the field improve water penetration. The more fibrous and less easily decomposable crop residues are more suitable for mitigating minor infiltration problems. Fibrous organic materials keep the soil porous by maintaining open voids and channels. The use of crop residues forming polysaccharides as a cementing agent is most effective where they are incorporated only in the upper few centimetres.

Incorporation of easily decomposable organic matter and incorporation deeper into the soil profile do not generally help reduce infiltration problems although they do improve soil structure by producing polysaccharides that promote soil aggregation and enhance soil fertility. Another important aspect of the decomposition of organic matter is the production of carbon dioxide, which in turn increases the solubility of lime.

Cultivation Practices

Cultivation is usually done for weed control or soil aeration purposes rather than to improve infiltration. However, where infiltration problems are severe, cultivation or tillage are helpful as they roughen the soil surface, which slows down the flow of water, so increasing the time during which the water can infiltrate. Cultivation is only a temporary solution. After one or two irrigations, another cultivation may be needed. Moreover, the construction of (broad) beds may help mitigate the ill effects of standing water as it prevents direct contact of the plants with the water. Research from Pakistan has shown that cotton in particular benefits from planting on broad beds.

Maintaining Favourable Levels of Ions and Trace Elements

High concentrations of trace elements in soil, ground and drainage water can occur in association with high salinity and can be affected by the same processes. However, in some places they may also occur independently of salinity. In examining ways to control levels of ions and trace elements in the rootzone, it is necessary to understand the processes that affect their mobility.

Deverel and Fujii provide a framework for evaluating concentrations of trace elements in soil and shallow groundwater. The two processes that largely control the mobility of trace elements in the soil water are:

i) adsorption and desorption reactions; and

ii) solid-phase precipitation and dissolution processes.

These processes are influenced by changes in pH, redox state and reactions, chemical composition and solid-phase structural changes at the atomic level.

There are no well-tested and simple models for estimating changes in trace element concentration as a result of irrigation and drainage water management. Nor have irrigation water quality criteria for trace elements been established. However, guidelines have been developed for trace elements based on results from sand, solution and pot trials, field trials with chemicals and laboratory studies of chemical reactions.

Selenium

Albasel et al. reviewed the quality criteria for trace elements in irrigation water compiled by the National Academy of Engineering in 1973. The recommendation for selenium was that the concentrations in irrigation water should not exceed 20 mg/litre. The guideline was recommended for all irrigation water on any land without consideration for soil texture, pH, plant species, climate and other water characteristics such as sulphate concentration. In the San Joaquin Valley, specific conditions indicated that a review of this guideline was required:

i) selenium in the drainage water is in the form of selenate, which is not readily adsorbed onto soil particles and is thus readily leached;

ii) water containing high selenium concentrations also has a high total salinity content and thus requires high leaching fractions to prevent salinity build up in the soil profile; and

iii) water containing selenium has high concentrations of SO_4^- that greatly inhibit plant uptake of selenium. Albasel et al. used the concentration fractions for various LFs to convert the concentration of selenium in the irrigation water (Se_I) into selenium concentration in the soil solution (Se_{ss}).

Research with alfalfa, which is the most sensitive crop to selenium accumulation in relation to the use of the harvested product, showed that Se_{ssm} is 250 mg/litre without exceeding the Sehp of 4-5 mg/kg if the saline irrigation water is dominated by sulphate. Assuming that the *LF* would be 0.2 or more, the Se_i could be 100 mg/litre. Where other crops are grown on soils different from those in the San Joaquin Valley, new guidelines will be necessary. These should be based on soil and water properties and the maximum Se_{hp} specific for those crops and the use of the harvested product.

Boron

Boron is an essential micronutrient for plants but it is toxic at concentrations only slightly above deficiency. The range of boron tolerance varies widely among crop plants. Salt sensitive crops such as citrus, fruit and nut trees are sensitive to boron while salt tolerant crops such as cotton, sugar beet and Sudan grass tolerate higher levels of boron.

Leaching of soil boron is more difficult than soluble salts such as chloride. This is because of the slow dissolution of boron minerals and desorption of boron adsorbed to oxides of iron and aluminium in the soil. Hoffman established a relationship to calculate the relative decrease of soluble boron in soils during reclamation:

$$\left(\frac{B_{sst}}{B_{ss0}}\right)\left(\frac{D_L}{D_S}\right) = 0.6 \qquad (1)$$

where:

B_{sst} = desired boron concentration in the soil solution (mg/litre);

B_{ss0} = initial boron concentration in the soil solution (mg/litre);

DL = depth of leaching water (mm); and

Ds = depth of soil to be reclaimed (mm).

Keren et al. reported that native soil boron is more difficult to leach than boron accumulated from irrigation with boron-rich water. Where steady-state conditions exist between boron adsorbed and boron in the soil solution (B_{ss}), then the input and output of boron from the rootzone, and thus Bss, is related to the boron concentration in the irrigation water (*BI*) and the *LF*. To establish the relationship between *BI* and B_{ss}, a water uptake pattern of 40-30-20-10 percent from the first to the fourth quarter of the rootzone is assumed.

Reuse in Conventional Crop Production

Drainage water of sufficiently good quality might be used directly for crop production. Otherwise, drainage water can be reused in conjunction with freshwater resources. Conjunctive use involves blending drainage water with freshwater. Alternatively, drainage water can be used cyclically with freshwater being applied separately. In cyclic use, the two water sources can be rotated within the cropping season (intraseasonal cyclic use), or the two water resources can be used separately over the seasons for different crops (interseasonal cyclic use).

The choice of a certain reuse option depends largely on: drainage water quality; crop tolerance to salinity; and availability of freshwater resources. The quantity and time of availability of drainage water is of major importance. For example, where reuse takes place in an irrigation system in which water is distributed on a rotational basis, the probable mode of reuse is either direct or cyclic.

Direct Use

The direct use of drainage water is implemented mainly at farm level, whereby the drainage water is not mixed with freshwater resources. Research results from India, Pakistan, Central Asia and Egypt show that drainage water can be used directly for irrigation purposes without severe crop yield reductions where the salinity of the drainage water does not exceed the threshold salinity value for the crops grown and good drainage conditions exist. As crops are often more sensitive to salinity during the initial growth stages, research in India has revealed the importance of pre-irrigation with good quality irrigation water. Higher crop yields were attained when freshwater pre-irrigation was applied with only drainage water being applied thereafter. Under these conditions, drainage water with salinity levels exceeding the threshold value could be used whilst maintaining acceptable crop yields.

Conjunctive Use

Blending: Where drainage water salinity exceeds the threshold values for optimal crop production, it can be mixed with other water resources to create a mixture of acceptable quality for the prevailing cropping patterns.

Where reuse takes place by mixing drainage water from main drains with surface water in main irrigation canals, the most salt sensitive crop determines the final water quality. For example, according to the regulations for the Nile Delta, the maximum salinity of the blended water is rather low to ensure optimal production of the major crops grown, i.e. cotton, maize, wheat, rice and berseem. Maize and berseem are the most salt sensitive crops with an ECe threshold of 1.7 and 1.5 dS/m, respectively. To ensure potential maximum yield production of these two crops and assuming a concentration factor of two between the EC of the irrigation water and the ECe, the maximum allowable salinity of the irrigation water is about 0.8-0.9 dS/m.

Table 1. Drainage water quality criteria for irrigation purposes in the Nile Delta, Egypt

Salinity of drainage water (dS/m)	Restriction on use for irrigation
< 1.0	used directly for irrigation
1.0 - 2.3	Mixed with canal water at ratio 1:1
2.3 - 4.6	Mixed with canal water at ratio 1:2 or 1:3
> 4.6	Not used for irrigation

Where mixing takes place at the farm level, the salinity of the blended water can be adjusted towards the salt tolerance of individual crops. Table 2 shows an experiment from India, where blended water with different levels of salinity was used to cultivate wheat.

Table 2. Effect of diluted drainage water on wheat yield

ECI, dS/m	Relative Yield (%)	
	All irrigations	Post-plant irrigations
0.5 (canal water)	100	100
3.0	90.2	-
6.0	80.4	95.8
9.0	72.5	90.3
12.0	56.4	83.7
18.0	-	78.0

This experiment shows that for the case of all irrigations applied with a blended water mixture having a maximum salinity of 3 dS/m, a potential crop yield of 90 percent of the maximum yield was obtained. According to the crop tolerance data published by Maas and Grattan and assuming a concentration factor of 1.5 between water salinity and soil salinity, the maximum potential yield could be attained. The difference between theory and the actual field situation might be caused by a discrepancy in the actual concentration factor, which is the inverse of the LF. Moreover, the actual yields depend on many factors including physical, climate and farm management practices. In general, where all irrigations use the blended water, the maximum allowable salinity depends on the threshold value of the crops. Where the pre-irrigation uses freshwater, the blended water could have a salinity level exceeding the threshold value without affecting yields. The experiment from India shows that water used in post-plant irrigations with a salinity level of three times (ECI = 9 dS/m) the maximum allowable salinity level when only blended water is used (ECI = 3 dS/m) results in 90 percent yields.

The leaching of salts is necessary to maintain a favourable salinity balance in the rootzone. The procedures presented above might be used for this purpose. On the other hand, if higher salinity levels are tolerated towards the end of the growing period or if a more salt sensitive crop is grown after a more salt tolerant crop, the salinity levels have to be reduced sufficiently low in order not to interfere with the growth of the next crop.

Cyclic Use

Cyclic use, also known as sequential application or rotational mode, is a technique that facilitates the conjunctive use of freshwater and saline drainage effluent. In this mode, saline drainage water replaces canal water in a predetermined sequence or cycle. Cyclic use is an option for where the salinity of the drainage water exceeds the salinity threshold value of the desired crop. A condition for cyclic use is that two different water sources can be applied to the field separately. Therefore, it is not normally applied at irrigation-scheme level but at a tertiary or farm level. In India and Pakistan, the canal irrigation water is delivered on a rotational basis to the watercourses (tertiary canal) and individual farms. This offers considerable potential for cyclic use on tertiary or farm level. Modelling and field studies have demonstrated the feasibility of the cyclic reuse strategy.

The cyclic use of drainage water can be either intraseasonal or interseasonal. The latter mode of cyclic use follows the same principles for each cropping season as the direct use of drainage water.In intraseasonal cyclic use, the strategy implies that non-saline water is used for salt sensitive cropping stages and saline water when the salt tolerance of the plant increases. Such experiments have been carried out in India.

Saline drainage water (ECI of 10.5-15.0 dS/m) was combined with canal water for use in a pearl millet/ sorghum-wheat rotation. In the experiment, canal water was used for pre-plant irrigations and thereafter four irrigations each of 50 mm depth were applied as per the planned modes to irrigate wheat. Pearl millet and sorghum were given pre-plant irrigation and thereafter did not receive further irrigation except by monsoon rains during the growth period. The results of the experiment support the cyclic use strategy.

No significant yield losses occurred in wheat when saline drainage water was substituted in alternate sequences (canal water - drainage water, or drainage water - canal water), or when the first two irrigations were with canal water and the remaining two irrigations were with drainage water, or when the first two irrigations were with saline drainage water and the remaining two irrigations were with canal water.

Intraseasonal cyclic use offers considerable potential as not all crops tolerate salinity equally well at different stages of their growth. Most crops are sensitive to salinity during emergence and early development. Moreover, the flowering stage is also critical. Tolerance to salinity generally increases with the age of the crop. An exception is the salinity tolerance of mustard during the flowering to reproductive development stage.The long-term sustainability of this drainage water management option entails devoting sufficient to maintaining a favourable salt balance and to preventing a buildup of trace elements in the rootzone to levels toxic to plant growth.

Cyclic use also requires attention to soil degradation as a result of using sodic water. A high exchangeable sodium percentage on the soil exchange complex does not normally lead to soil degradation if it is compensated by a high soil moisture salinity to suppress the extent of the so-called diffuse double layer. However, upon irrigation with low saline irrigation water or rainfall, the diffuse double layer swells resulting in soil dispersion. Adding sufficient soil or preferably water amendments to compensate for the high sodium to calcium and magnesium content in the saline irrigation water can prevent these problems.

Crop Substitution and Reuse

Crops differ significantly in their tolerance to concentrations of soluble salts in the rootzone. The difference between the tolerance of the least and the most sensitive crops may be tenfold. A number of salt tolerant crop plants are available for greater use of saline drainage effluent. Raising the extent of the salinity limits through selecting more salt tolerant crops enables greater use of saline drainage effluent and reduces the need for leaching and drainage.

Reuse for Irrigation of Salt Tolerant Plants and Halophytes

Where the irrigation water is too saline to grow conventional agricultural crops, irrigation of halophytes might be considered. The maximum amount and kind of salt that salt tolerant plants and halophytes can tolerate vary among species and varieties. Halophytes have a special feature as their growth is improved at low to moderate salinity levels.

In contrast, salt tolerant crops have maximum growths up to a threshold salinity level after which growth is reduced. Salt tolerant plants and halophytes have been grown successfully in many places in the world to produce fuel, fodder and to a lesser extent food. Institutes in Australia have gathered useful salt tolerance data on a large number of trees and shrubs species. Growing salt tolerant plants and halophytes under saline conditions requires management of the salt balance in the rootzone. Where natural drainage is insufficient, artificial drainage is required to remove the leaching water.

Fuel: Many people in developing countries rely on wood for cooking and heating. As agricultural land is required to feed growing populations, it is unlikely that good quality agricultural land will be used for fuel production. Salt tolerant trees and shrubs can be grown for fuel production and building materials using saline water and marginal lands. Among the promising tree species for these purposes are Prosopis, Eucalyptus, Casuarina, Rizophora, Melaleunca, Tamarix and Acacia.

Fodder: Pasture improvement programmes in salt-affected regions throughout the world have used halophytes and salt tolerant shrubs and grass species. Trees and shrubs can be valuable complement to grasslands. They can serve as a nutrient pump and lower saline shallow groundwater tables. They are less susceptible to moisture deficits and temperature changes than grasses. They might also provide valuable complementary animal food or fuelwood. Salt tolerant grasses, shrubs and trees with potential for fodder use include:

Other products: Salt tolerant plants can also produce other economically important materials, e.g. essential oils, gums, oils, resins, pulp and fibre. Moreover, salt tolerant plants can be used for landscape and ornamental purposes and irrigated with saline water, thereby conserving freshwater for other purposes.

Reclamation of Salt-affected Land

Sodic soils often have low hydraulic conductivity as a result of the high sodium percentage on the soil exchange complex. The reclamation of sodic soils requires that a divalent solute (mainly calcium) pass through the soil profile, replacing exchangeable sodium and leaching the desorbed sodium ions from the rootzone. Therefore, the rate at which sodic soils can be reclaimed depends on the water flow through the soil and the calcium concentration of the soil solution.

The application of leaching water with a high electrolyte concentration promotes flocculation of the soils and thus improves soil permeability. This expedites the reclamation process. Amendments need to be added in order to replace sodium with calcium ions on the soil exchange complex. Over time, less-saline water needs to replace the saline leaching water in order to lower the salinity levels sufficiently to establish a crop.

The use of saline drainage water to reclaim salt-affected soils is not a permanent solution for reducing drainage effluent disposal volumes. It is only a substitute for the use of good quality irrigation water for reclamation purposes.

Drainage Water Disposal

Drainage water management is normally concerned with reducing the amount of drainage water and with managing its disposal. However, this aim is more complex than it appears. Drainage is practised to maintain aeration in waterlogged rootzone and/or to leach excess soil salinity to sustain agricultural production. The drainage water generated must then be managed for reuse purposes where it is of suitable quality and finally discharged or disposed of.

The discharge of drainage waters in watercourses may have impacts ranging from beneficial to deleterious. The disposal of drainage water into wetlands, lakes, rivers and coastal waters entails considerations about the quantity and quality allowable and indeed sometimes required to maintain desirable ecological conditions and functions of that given water body.

To sustain irrigated agriculture, the maintenance of favourable salinity levels is the major concern in assessing the minimum drainage disposal requirements. When aimed at maximising source reduction or reuse of drainage water, minimum leaching is required to maintain favourable salt balances in the rootzone. The minimum LF depends on the salinity of the irrigation water and the salt tolerances of the crops grown.

Minimum disposal requirements also depend on the requirements of downstream water uses. For example, many inland fisheries are currently threatened because of increasing water pollution, degradation of aquatic habitats and excessive water abstraction. With the increase in environmental awareness, there is increasing pressure to preserve sufficient water for aquatic environments, habitats and biodiversity. For example, wetlands are considered to be among the world's most valuable ecological sites. They form the habitat for many species of plants and animals.

Wetlands perform various other vital functions, e.g. water storage, flood mitigation and water purification. Wetlands also provide economic benefits, of which fisheries, recreation and tourism are among the most important. These functions, values and attributes can only be maintained if the ecological processes of wetlands continue functioning. If the ecological processes in the wetlands depend on drainage water inflow and if the functions of the wetlands are to be sustained, the minimum drainage discharge will depend on the quality and quantity of drainage water required for maintaining the ecological processes.

Depending on the location, hydrology and topography of the drainage basin (and the ecology and environmental conditions of receiving water bodies), drainage water might be disposed to open surface water bodies, e.g. rivers, lakes, outfall drains, and oceans. The oceans are often regarded as the safest and the final disposal site for agricultural drainage water. This is true unless drainage water is contaminated with sediments, nutrients and other pollutants and the disposal site is in the vicinity of fragile coastal ecosystems such as mangroves and coral reefs. Therefore, at the point of discharge, pollution may be of much concern.

In general, oceans have significant dilution or assimilative capacity. However, this is also limited in many cases especially in enclosed and semi-enclosed seas. Inland drainage water disposal to freshwater bodies such as lakes and rivers requires care. Rivers are normally used for different water use purposes requiring certain qualities of water, while the accumulation of salts and other pollutants in freshwater lakes threatens ecosystem functions and aquatic life.

Disposal into rivers or oceans is not always possible. In closed drainage basins, alternative disposal options need to be sought. The options for closed hydrologic basins include evaporation ponds, deep-well injection and integrated drainage management systems. Drainage water management in closed basins presents numerous environmental and water quality challenges.

Disposal in wetlands offers much potential but at the same time there are also accompanying constraints. Wetlands might be either natural or specially constructed for drainage water reuse. In the former case, the maintenance of ecological functions depends on the quality and quantity of drainage water inflow.

Disposal in Freshwater Bodies

The main aim of safe disposal in freshwater bodies such as lakes and rivers is to protect beneficial downstream water uses. Rivers and lakes are normally multifunctional (municipal drinking-water, industrial water, fishing, recreation and agriculture) and have an intrinsic ecological value. Furthermore, rivers feed lakes, floodplains, wetlands, estuaries and bays. To protect these functions, it is necessary to determine the assimilative capacity of the river and, where necessary, of connected or receiving ecosystems. It is also necessary to identify the constituents of concern in the drainage water in order to determine its discharge requirements. Moreover, the effects on sediments, riparian habitats and floodplains warrant consideration.

The discharge requirements should specify the maximum allowable concentration of each constituent of concern and the volume of drainage water that will be acceptable. Concentration is the water quality parameter normally used in drinking-water standards or for the health of aquatic animals and plants. However, placing only concentration limits on discharge might encourage dilution and inefficient water use.

Where both concentration and load limits (product of water concentration and water volume) are enforced, they tend to promote efficient water use. For example, in the San Joaquin Valley, the discharge limitations into the San Joaquin River include limits on concentration of specific constituents and their loads. The concentration limitations include limits on salinity, boron, selenium and molybdenum in order to protect downstream water quality for domestic and agricultural uses and to safeguard the environment. To discourage dilution or inefficient water usage, load limits for these constituents are enforced for the discharge of irrigation return flows into the San Joaquin River.

The assimilative capacity of the receiving water body varies from place to place and from time to time depending on numerous local conditions. These include climate and physical conditions and the upstream uses. The River Yamuna in Haryana, India, provides an example of seasonal differences in disposal opportunities. During the dry winter season, November-March, the flow of the River Yamuna is less than 50 m^3/s. From July to September, during the monsoon season, the flow exceeds 1 000 m^3/s with a peak of 1 150 m^3/s in August. The salinity of river water during monsoon is less than 0.2 dS/m. The river's high flow and low salinity during the monsoon period provides an opportunity for the disposal of saline effluents.

As the water table shows a marked rise during the monsoon season, major pumping for drainage takes place between July and September. Table 3 shows projections of the amounts of subsurface drainage water that could be disposed into the River Yamuna. The criterion used was that the resultant salinity of the water in the river after mixing should be less than 0.75 dS/m. A significant drainable area with salinity problems can use the River Yamuna as a possible drainage outlet between July and September.

Table 3. Allowable subsurface drainage discharge and drainable area into the River Yamuna

Month	Allowable discharge(m^3/s)		Drainable area(ha)	
	Effluent salinity (dS/m)			
	6	10	6	10
June	0.9	0.5	5 000	3 000
July	25.4	14.4	146 000	83 000
August	47.6	27.0	274 000	156 000
September	6.5	3.7	37 000	21 000
October	3.0	1.7	17 000	10 000

Drainage from the Grassland subarea in the San Joaquin Valley provides another example of the possibilities and constraints of river disposal. In the recent past, irrigated agriculture discharged its irrigation return flow into the North and South Grasslands District, an entity devoted to wetland habitat for private duck clubs and birdwatching as well as pastureland for grazing animals. Since the discovery of selenium poisoning of waterbirds at Kesterson Reservoir from subsurface drainage waters conveyed from the Westlands Water District, drainage from the problem area is no longer discharged into the Grasslands Water District. Instead, it is discharged directly into the San Joaquin River at Mud Slough via the San Luis Drain and a bypass.

Although their flows are minor, Salt Slough and Mud Slough are major contributors of TDS and selenium in the lower reaches of the San Joaquin River. Thus, the Central

Valley Regional Water Quality Control Board (CVRWQCB) has placed monthly waste discharge requirements for Salt Slough and Mud Slough. The water quality objectives were established to protect downstream water uses as well as export water to southern California. The salinity water quality objective for Vernalis, a benchmark station on the river, is 1 dS/m for a 30-day running average from 1 September to 30 March, and 0.7 dS/m for a 30-day running average from 1 April to 31 August.

The boron objective is placed from Sack Dam to Merced River, the reach above and below Salt Slough and Mud Slough, at 2.0 mg/litre monthly mean for the year with a maximum of 5.8 mg/litre for a given month. The selenium water quality objective is placed on Salt Slough and Mud Slough at 5 mg/litre, 4-day average, to protect aquatic biota and a maximum annual load of 3 632 kg to control mass emission. Exceeding the specified allowable loads of salt and selenium results in monetary fines to the dischargers. The selenium load is the most difficult for irrigated agriculture to meet.

Generally, rivers continually cleanse themselves. In most cases, lakes do not have this capacity as they may not have an outlet, or flow volume from the lake is limited. The disposal of drainage water into lakes might cause substantial long-term problems. Thus, it involves special considerations. There are numerous examples from all over the world where disposal of agricultural drainage water into freshwater lakes, often in combination with reduced freshwater inflow, has added to the development of immense and catastrophic environmental problems. Examples include Lake Chapala in Mexico, Lake Manchhar in Pakistan, Lake Biwa in Japan, and other lakes where significant effort is required to halt environmental degradation.

Disposal into Evaporation Ponds

In inland drainage basins without an adequate outlet to a river, lake or sea, or where disposal restrictions prevent discharge into rivers and lakes, one of the few options is disposal of drainage effluent into constructed evaporation ponds or natural depressions. Discharge into natural depressions has been practised for centuries. The impounded water is dissipated by evaporation, transpiration and seepage losses. Disposal into constructed ponds is practised worldwide. The planning and design of evaporation ponds have to take account of numerous environmental problems. These include:

- — waterlogging and salinisation problems in adjacent areas resulting from excessive seepage losses;
- — salt-dust and sprays to areas downwind of the pond surface during dry and windy periods, which might damage vegetation and affect the health of humans and animals; and
- — concentration of trace elements that might become toxic to fish and waterbirds because of bioaccumulation in the aquatic food chain.

The following examples from Pakistan, the United States of America and Australia highlight the opportunities and constraints of evaporation ponds.

In Pakistan, evaporation ponds have been constructed under the Salinity Control and Reclamation Project (SCARP) VI for the disposal of the highly saline drainage effluent from the irrigated areas on the fringes of the desert located 500-800 km from the sea. SCARP VI has employed vertical drainage to lower and maintain the water table at a depth of 2.3 m over an area of 152 000 ha. The area is underlain by groundwater with an EC ranging from 15 to 34 dS/m. A total of 514 tubewells, ranging in capacity from 30 to 90 litres/s, have been installed to pump out 600 million m^3 annually. This water is discharged into open surface drains for conveyance to the evaporation ponds for final disposal. Allowing for conveyance losses, the drainable surplus to be disposed of is an estimated 540 million m^3. It was anticipated that the resultant salinity in the drainage water would be of the order of 30 dS/m.

The design of the evaporation ponds assumed a percolation rate of 280 mm per year (half of the estimated value) and an evaporation rate of 1 700 mm per year (against the observed value of 1 800 mm). The surface area of the ponds was estimated to be 27 300 ha. The evaporation ponds were developed in a desert area on the fringes of the Indus Plain just beyond the canal command area of the project. The pond area is characterised by interdunal depressions (with highly sodic soils), between longitudinal sand dunes 4-9 m high. The underlying groundwater is highly brackish.

For the development of the evaporation ponds, dykes were provided across the saddles and channels cut across the dunes to form a series of interconnected ponds. The ponds that have been developed so far have a surface area of 13 350 ha and a distance of 1-3 km separates their edges from the irrigated area. The drainage effluent is discharged into the evaporation ponds by gravity flow from one drainage system and pumped from another drainage system further downslope.

The operation of the ponds started in 1989 and by May 1998 it was reported that water had spread over a pond surface area of 4 200 ha. The period of operation has been too short to establish a state of equilibrium. However, soon after ponding started some irrigated areas close to the ponds were severely affected by waterlogging. This could be ascribed to significant seepage losses from the ponds generating a zone of high groundwater, obstructing or retarding the natural subsurface drainage from the irrigated lands. The seepage affected area had grown to about 4 000 ha in 1993, even though the tubewell pumping was effective in lowering the water table to 3 m in parts of the drained area.

In California, between 1972 and 1985, 28 evaporation ponds were constructed covering an area of about 2 800 ha receiving about 39 million m^3 annually of subsurface drainage from 22 700 ha of tile drained fields. Basinwide, the pond area is about 12 percent of the total area drained. Most of the ponds are located in the Tulare subarea, a closed basin in the San Joaquin Valley. The salt concentration in the waters discharged into these ponds ranged from 6 to 70 dS/m with an annual salt load of 0.88 million tonnes, about 25 percent of the annual salt load accumulating in the more than 0.9 million ha of cropland.

The concentration range of selenium, the principal constituent of concern, is from 1 to over 600 ppb. A selenium concentration of 2 ppb is considered the upper limit for aquatic life in ponds as selenium tends to bioaccumulate 1 000- to 2 500-fold in the aquatic food chain. Selenium is toxic to waterbirds as it substitutes for sulphur in essential amino acids resulting in embryo deformities and reduced reproduction rates. Ten ponds are active and managed by seven operators. The other ponds have been voluntarily deactivated due to the high costs of meeting the waste discharge requirements and mitigation measures for bird toxicity. The state/regional water quality regulatory agency ordered the closure of several ponds because of the unacceptable toxic effects of selenium on waterbirds from selenium present in the impounded drainage waters.

The waste discharge requirements for pond disposal define the requirements and compliance schedules designed to discourage wildlife use of evaporation ponds and/or provide mitigation and compensation measures to offset adverse effects on birds.

The evaporation ponds are constructed by excavating the soil to form berms with side-slopes of at least 3:1 (h:v). The pond bottoms are unlined but compacted, the pond environs are kept free of vegetation, and a minimum water depth is maintained at 60 cm. Migratory waterbirds attracted to the pond site are to be scared off, and bird diseases kept under control.

Compliance monitoring includes seasonal drainage water and sediment selenium concentrations, and biological monitoring of birds for abundance and symptoms of toxicity. When adverse selenium impacts are noted, off-site mitigation measures must be implemented with either compensation or alternative habitats as guided by approved protocols and risk analysis methods.

Compensation habitats are constructed to compensate for unavoidable migratory bird losses by providing a wetland habitat safe from selenium and predators. Alternative habitats are year-round freshwater habitats immediately adjacent to contaminated ponds in order to provide dietary dilution to selenium exposure. In spite of the stringent waste discharge requirements and associated high costs, farmers and districts in the Tulare subarea use evaporation ponds for drainage water disposal basins. This is because there are no opportunities for off-site discharge into the San Joaquin River or other designated sinks for drainage waters. Evaporation ponds are the only economic means of disposal of saline drainage in the Tulare-Kern subareas.

In Australia, the Salinity Drainage Strategy of the Murray-Darling Basin Commission imposes constraints on the amount of river disposal. Export of saline drainage water via pipelines to the sea has been considered but studies show that this option is relatively uneconomic compared to other disposal options. Saline disposal basins have been an important measure for the disposal of drainage effluent and will remain so. In the Murray-Darling Basin in 1995, 107 basins were actively used with a total area of more than 15 900 ha, a total storage capacity of more than 113 million m^3 and an annual disposal volume of more than 210 million m^3/year.

The use of regional-scale basins was a common approach in the Murray-Darling Basin. Regional basins were developed on the most convenient sites from an engineering standpoint. This sometimes had detrimental environmental, socio-economic and aesthetic impacts. These concerns together with the new opinion that beneficiaries of irrigation should be responsible for their own drainage management led to the development and use of local-scale basins.

These basins can be in the form of on-farm basins which are privately owned and located on individual properties. They can also be privately or authority-owned community basins shared by a small group of properties. The choice between on-farm or community basins should consider physical, environmental and sociopolitical issues as well as costs. Economic analysis has shown that there is generally little difference in cost between the two options. Therefore, environmental and social considerations should outweigh the negligible economic differences in the choice between on-farm and community basins in the Murray-Darling Basin.

Design of evaporation ponds

The first step in the design of an evaporation pond is to determine the volume of drainage water in need of disposal. The volume of drainage water depends on the land use, climate, irrigation practice and type of drainage system. The planning stage needs to include an assessment of the salinity of the drainage water as well as a full analysis for trace elements, nutrients and pesticides to prevent possible accumulation to toxic levels.

Trace elements in drainage water disposed into the evaporation ponds include arsenic, boron, molybdenum, selenium, uranium and vanadium. The trace element of most concern in California for impounded drainage waters is selenium. The influent selenium is subjected to a number of reactions such as evapoconcentration, seepage losses, reduction of oxidised species of selenate (Se+6) and selenite (Se+4) to selenide (Se-2) as well as elemental selenium (Se0), methylation by microbes and plants and volatilisation as dimethylselenide (DMSe), and adsorption of selenite to pond bottoms. Approximately 60 percent of the mass of selenium influent appears to remain in the ponded water, the balance being immobilised or lost by the above mechanisms.

The hydrologic balance of evaporation ponds is relatively simple. The inputs into the pond include drainage water from cropped land, rainfall and, where applicable, water pumped into the pond from a perimeter drain installed to intercept seepage water from the pond. The outputs consist of evaporation, non-recovered seepage losses and transpiration from aquatic vegetation. An example of the hydrology of a typical pond system in California is: 1.91 m/year drain influent; 0.022 m/year precipitation; 1.87 m/year evaporation; 0.39 m/year seepage; and 0.46 m/year influent from the perimeter interceptor drain that collects both pond seepage losses and shallow groundwater.

The required pond area can be calculated on the basis of this hydrologic balance and depending on the depth of water layer that is to be maintained in the pond. When the salinity of the water in the evaporation ponds increases, the evaporation rate from the water surface decreases as a result of a decrease in water vapour pressure. The design of the pond area needs to take this aspect into account. Evaporation rates in ponds have been measured. The pond evaporation can be estimated using an empirical correction factor (Y) and ETo.

$$E = Y(Eto) \qquad (2)$$

in which Y = 1.3234 - 0.0066 EC for water between EC 14 and 60 dS/m.

Seepage losses from evaporation ponds depend on soil, geohydrological and topographical conditions. Site selection is very important for preventing excess seepage losses. Depending on the hazards that excessive seepage might occur and depending on the environmental regulations, lining and a seepage collection system may be required in order to ensure that there will be no contamination of groundwater, and to prevent waterlogging in adjacent areas.

The availability of materials should determine the type of lining. Research from Australia has shown that a small amount of controlled seepage of around 0.5-1.0 mm/d is required to maintain evaporative disposal capacity, as evaporation rates decrease substantially under hypersaline conditions. This implies that basins should be located preferably within the drained area. Where the basin is sited outside the limits of the drainage system, it should be located within a specific salt containment area equipped with effective interception and recycling works.

For those pond systems deemed to be hazardous to waterbirds, pond operators are required to establish alternative and compensation habitats. Alternative habitats using freshwater are built within the functional landscape of evaporation ponds to provide year-round habitat to dilute the diet to the feeding birds. Compensation habitats are located outside the functional landscape of the evaporation ponds during the breeding season.

Injection into Deep Aquifers

Disposal by injection into deep aquifers is a process in which drainage water is injected into a well for placement into a porous geologic formation below the soil surface. The oil and gas industries use deep-well injection to dispose of waste brine.

In California, the United States of America, deep-well injection technology has been used for over 60 years for the disposal of oil field brines. The technology could potentially be applied for the disposal of agricultural drainage water provided the conditions of the receiving geologic formation are adequate, there are no environmental hazards and the costs are not prohibitive. The main environmental concern is leakage from pipes and confining aquifers. Drainage effluent can contaminate freshwater zones through leakage.

The drainage and aquifer water should be compatible, as the mixture should not produce precipitates, which clog the wells. Furthermore, the receiving geologic formation should have sufficient porosity and thickness to receive the injected water. If water containing nitrate is injected into aquifers containing organic matter and ferrous iron, the growth of nitrate reducing bacteria might clog the pores of the receiving formation as they accumulate.

The Westlands Water District, California, the United States of America, carried out a prototype deep-well injection to dispose of up to 4 000 m3/d of drainage water. The drainage water was to be in injected into shale and sand formations 1 554 and 2 164 m, respectively, beneath the ground surface. The well was drilled to a total depth of 2 469 m at a cost of about US$1 million. The casing was perforated with 13 perforations per metre from depths of 2 245 to 2 344 m and from 2 411 to 2 414 m in the Martinez geologic formation for a total length of perforations of 102 m.

Following the recovery of some natural formation water samples, an injection test was conducted. The fluid injected was irrigation water filtered through a 0.5-micron filter and treated with 2-percent potassium chloride and a chlorine biocide. The filtered and treated water was injected through a 4.75-cm tubing at a rate of 12 litres/s for a total of 175 000 litres of water injected. Then, a 48-hour pressure fall-off test was conducted that revealed the permeability of the geologic formation was 12 mm/d. This permeability value is too low to achieve the desired injection rate of 44 litre/s. In spite of the filtration and chlorination of the injected water, plugging of conducting pores occurred in the shale.

The United States Environmental Protection Agency rejected a request for permission to conduct a second injection test in the sandy, more permeable formation above the shale. The Westlands Water District abandoned this method of disposal.

Treatment of Drainage Effluent

Treating drainage water is normally one of the last drainage water management options to be considered. This is due to the high costs involved and to uncertainty about the treatment level achievable. The treatment of drainage water should be considered where all other drainage water management measures fail to guarantee safe disposal or where it is financially attractive. For subsurface drainage water containing very high levels of salinity, selenium and other trace elements, the treatment objectives are:

i) reduce salts and toxic constituents below hazardous levels;

ii) meet agricultural water management goals;

iii) meet water quality objectives in surface waters; and

iv) reduce constituent levels below risk levels for wildlife.

The treatment of agricultural drainage water presents a challenge due to the complex chemical characteristics of most drainage waters. Table 4 details the average chemical quality of subsurface drainage waters disposed into Kesterson Reservoir in the San Joaquin Valley as well as those disposed into evaporation ponds.

The drainage waters are saline and of the NaCl-Na_2SO_4-type water. The waters conveyed by the San Luis Drain came from a single site in Westlands Water District in contrast to the evaporation pond waters that came from 27 sites.

There are numerous wastewater treatment processes for industrial and urban wastewater and for the preparation of drinking-water. Many of them offer potential for the treatment of agricultural drainage water.

Table 4. Average composition of agricultural tile drainage water in the San Luis Drain (drainage waters disposed into evaporation basins in parenthesis)

Constituent	Concentration (ppm)	Constituent	Concentration (ppb)
Sodium	2 230	Boron	14 400 (25 000)
Calcium	554	Selenium	325 (16)
Magnesium	270	Arsenic	1 (101)
Potassium	6	Molybdenum	ND (2 817)
Alkalinity as $CaCO_3$	196	Uranium	ND (308)
Sulphate	4 730	Vanadium	ND (22)
Chloride	1 480	Strontium	6 400
Nitrate	48	Total chromium	19
Silica	37	Cadmium	<1
TDS	9 820 (31 000)	Copper	4
Suspended solids	11	Lead	3
Total organic carbon	10.2	Manganese	25
COD	32	Iron	110
BOD	3.2	Mercury	<0.1
		Nickel	14
		Zinc	33

Treatment processes for drainage water can be divided into processes that reduce the total salinity of the drainage water and processes that remove specific ions. Methods for the removal of trace elements can be biological, physical and chemical.

Desalinisation

There are numerous desalinisation processes including ion exchange, distillation, electrodialysis and reverse osmosis. Of these processes, reverse osmosis is considered to be the most promising for the treatment of agricultural drainage water mainly due to its comparatively low cost.

Reverse osmosis is a process capable of removing different contaminants including dissolved salts and organics. In reverse osmosis, a semi-permeable membrane separates water from dissolved salts and other suspended solids. Pressure is applied to the feed-water, forcing the water through the membrane leaving behind salts and suspended materials in a brine stream. The energy consumption of the process depends on the salt concentration of the feed-water and the salt concentration of the effluent. Depending on the quality of the water to be treated, pretreatment might be crucial to preventing fouling of the membrane.

The California Department of Water Resources conducted pilot-plant-scale reverse osmosis of saline drainage using cellulose acetate membranes. The bacterial and chemical fouling of the membrane was a major problem. The drainage water had to be treated with alum, and passed through a sedimentation pond and a chlorinated and filtration system.

In spite of this level of pretreatment, the membranes tended to foul due to the precipitation of gypsum and calcite. The drainage waters are saturated with respect to calcite and gypsum. This same chemical fouling problem is being faced by the Yuma desalting plant off the Colorado River using drainage waters from the Wellton-Mohawk irrigation project. The estimated cost of desalting is more than US$0.81/m^3, too expensive for irrigated agriculture but possibly affordable for municipalities with freshwater shortages. This cost does not include the management and disposal of the brine water. However, a potential exists for partially treating the average 10-dS/m-drainage water to about 2-3 dS/m for use by agriculture and wildlife.

6

Capacity Assessment in Agricultural Water Management

Capacity is not something that can be built through a series of carefully planned and executed activities that follow a clear and detailed plan or blueprint with specific timeframes and strict budgets. It is an organic process of growth and development involving experimentation and learning as it proceeds. Therefore, many people now speak of capacity development rather than capacity building, to emphasise that it is a process rather than a blueprint.

Capacity development is the sum of efforts needed to nurture, enhance and utilise the skills and capabilities of people and institutions at all levels — locally, nationally, regionally and internationally — so that they can better progress towards sustainable development. At the basic conceptual level, building capacity involves empowering people and organisations to solve their problems, rather than attempting to solve problems directly. When capacity development is successful, the result is more effective people and institutions better able to provide products and services on a sustainable basis.

Five strategic phases of capacity development can be identified. The first phase is an assessment to define present capacity within the system. It establishes the baseline and addresses the basic question—*where are we now?* The second phase looks ahead to the future desired state, the vision of what capacity is required in the future and asks the question—*where do we want to go?* The third phase compares the present situation and future desired state, identifies the capacity gaps and plans strategies and actions designed to fill these gaps and achieve the desired goals—*how do we get there?* The fourth phase is the implementation phase, fulfilling the strategies and undertaking the planned capacity development activities in order to meet the defined objectives—*what actions do we take?* The final phase is monitoring and evaluation to feed back experiences into the planning phase—*how do we stay there?* This is not a linear process, the phases are interlinked and overlap, they form a continuing cycle of development and change according to the prevailing circumstances.

Of the five phases, assessment of capacity needs, that is, establishing the existing and required capacities as well as identifying the gaps between both, is perhaps the least well developed, at the same time it is the most vital. Without a proper understanding of what currently exists and what is needed, there is a good chance that inappropriate measures and actions will be initiated. Most technical assistance programmes and projects focus on meeting capacity needs, without undertaking the analysis required to ensure the solution is the most suitable in the circumstances.

Process of Capacity Needs Assessment

A review of the literature and official publications indicates there is a considerable body of work available on capacity development, though very little has been published on approaches to assessing capacity needs. UNDP has published extensively on capacity development and, in particular, on assessing capacity needs. Their approach is simple in concept—first assess the existing capacity, then assess future capacity envisaged by answering the question—*where do we want to go?*—and from these two, identify the capacity gaps. Strategies can then be developed to fill the gaps.

UNDP Model for Needs Assessment

The UNDP guidelines introduce a note of realism by suggesting that putting this concept into practice is not a simple process. The first step of assessing existing capacity can usually produce much useful information. The second step, of looking at future capacity, is less sure as it largely depends on policies and strategies for future development and these may not always be as clear as they should be. This has a knock-on effect on the third step of assessing the gaps and hence on developing strategies to fill the gaps.

UNDP makes the case for capacity assessment as a structured and analytical process, whereby the various dimensions of capacity are assessed within the broader socio-economic environment, as well as evaluated for specific organisations and individuals. They indicate how assessments might be undertaken in different situations and how programmes and projects might be better designed to ensure ownership, sustainability and ultimately success. Special emphasis is given to using existing capacities as the basis for capacity development. The guidelines indicate that, as no two situations are alike, they require common sense and flexibility in their application, as well as a good understanding of the particular context in which they are used. The UNDP model for needs assessment based on the three levels of capacity (Figure 1) provides the basis for the first three phases of capacity development. These are assessing existing capacities, identifying possible future capacity, estimating the gaps and defining possible strategies. The capacity levels are expanded to include a number of sub-levels or dimensions that need to be considered in the assessment.

Enabling environment: Assessments at this level include the dimensions of policy, legal and regulatory framework, management and accountability, resources and processes (the inter-relationships between organisations) and how resources and information flow both formally and informally. For development initiatives that are national in context (e.g. governance or environmental programmes), all the dimensions would need to be examined. However, for initiatives at a sectoral level, the analysis would include only the relevant components.

Organisations: Many different organisations influence irrigated agriculture. They may be government departments, private sector operations or informal community-based associations. Traditionally capacity development and organisational strengthening focused almost entirely on human resources, processes and organisational structuring. Nevertheless, this tends to be too narrow a focus and additional dimensions need to be assessed including mission and strategy, culture, structure and competencies, processes (both internal and external), human resources, financial resources, information resources and infrastructure. It is also important to include interactions with other organisations and stakeholders within the wider enabling environment. In an increasingly complex world, it is important to form partnerships and to network with other organisations for synergy and complementarity in delivering services.

Dimensions of capacity Environment - Level I	Existing capacity	Possible future capacity	Estimated capacity gap	Possible strategies
Policy framework Legal and regulatory framework Management accountability Resources Processes and relationships				

Dimensions of capacity Organization - Level II	Existing capacity	Possible future capacity	Estimated capacity gap	Possible strategies
Strategic management Culture/structure Processes Human resources Resources - financial Resources - information Infrastructure Inter-relationships				

Dimensions of capacity Individual - Level III	Existing capacity	Possible future capacity	Estimated capacity gap	Possible strategies
Job skills and needs Professional development Access to information Performance/incentives Values/attitudes/motivation Relationships/interdependence Professional integrity Communication skills				

Figure 1. Matrix for assessing capacity

Individuals: Many capacity initiatives ultimately concentrate on the individual, either because this level is considered the most critical, or because it is the easiest to address. It considers the individual's capacity to function efficiently and effectively within an organisation and within the broader environment. Often, capacity assessments of individuals are based on an established job description or some other format that lays out the performance/skills requirements of the position and the individual filling that position. This is combined with a skills assessment of the individual.

The assessment is designed to assess individual capacity gaps so training and development plans can then be prepared to address them. Increasingly, the dimensions of accountability and incentives performance, values and ethics are becoming even more important at the level of individual capacity assessments. Strategies that stress continuous learning and professional development are also important.

Mapping capacity and finding the gaps: The rows represent the various dimensions of capacity. In most situations, an assessment will generate information on existing capacity. Moving to the next step of assessing future capacity may be less certain, information on this will only emerge gradually as answers to the question—*where do we want to be?*—emerge.

Existing capacity: A critical issue for undertaking an assessment is the entry point. Once an entry is made, movement can then take place from one level, enabling environment, organisation or individual, to another according to needs. Some information, gathered from the assessment at a certain level, might indicate that another level should be assessed. The entry point for national assessments will need to be at the level of enabling environment. An example might be a major initiative to improve the role and functioning of the legal framework (laws, legislation) as part of a water governance reform programme. Following an assessment of the broader dimensions of capacity, this could enable the assessment to 'zoom in' and examine the dimensions of specific organisations, such as the judiciary or the role of the Ministry of Water Resources and, as a result, also the role of individuals with organisations.

Future capacity: Assessing future capacity can be a difficult and time consuming step, as stakeholders may not always have a clear vision of what they would like to achieve. A realistic assessment of development objectives, resources and timing is required. A general consensus among all the stakeholders involved is also needed if the goals are to be achieved.

A common vision of the future enabling environment is an essential step to a more detailed assessment of the various dimensions at this level and of assessments at the lower levels of organisations and individuals. In practice, however, it may not always be possible to follow such a logical pathway. A process of iteration over an extended period is more likely as policy is developed and changed over time. Capacity is not static; it is a lengthy and continually developing and changing process.

Tools for Capacity Assessment

A range of tools and approaches exist for assessing capacity at the different levels.

Enabling Environment

At the level of the enabling environment, the two fundamental needs are to map and understand the policy environment and to assess the legal and regulatory framework. Once these are understood, the other issues to be considered at this level, such as accountability, resources, processes and relationships, can usually be identified and mapped with relative ease.

A number of approaches are possible for mapping the policy environment such as policy characteristics analysis, political mapping, or policy network mapping. All are based on simple unifying themes and can be presented as models or matrices.

— *Policy characteristics analysis*: assists in identifying the driving forces, decision-makers, costs and benefits, and those who bear the costs and benefits of policy-making in particular sectors.

— *Political mapping:* identifies stakeholders in policy-making, differentiating between external stakeholders, social groups, political parties and pressure groups, and distinguishing their different levels of support and influence, and

— *Policy network maps*: try to graphically present the linkages between different aspects and sectors in policy-making, in order to identify inconsistencies and constraints and to support so-called 'joined-up government'.

The legal and regulatory framework must also be assessed to see if it helps or hinders agricultural water management and to identify the gaps and inconsistencies. This requires an examination of existing laws, their extent and coverage and the accompanying regulations or regulatory systems, as well as attempts to assess their impacts through some form of regulatory impact assessment. Currently, a growing area of interest is water rights and their relationship with water laws. These also need to be considered within this general framework.

Both the policy processes and legal framework must take account of the management of water for agriculture within its wider context. For example, the Global Water Partnership has developed a 'toolbox' for integrated water resources management. The toolbox comprises brief references to many relevant issues at each level of the matrix. It is also useful for ensuring that the irrigation and drainage sector is considered within the wider framework of water resources, of which it plays such an important part.

Organisational Level

Assessment of existing capacity at the organisational level must take account of these organisations within their institutional context and consider their linkages and relationships with other relevant organisations. A first step in capacity assessment at the organisational level is therefore some form of stakeholder analysis.

The full participation of all those who have a stake in the outcomes of the assessment (the stakeholders) is now well recognised as an essential part of the process. These are the farmers and their organisations who are the principal beneficiaries of irrigation, the wider local community, local and national government organisations and private sector organisations, such as NGOs and commercial companies. Stakeholder analysis can be used to determine who is and should be involved, the nature of their involvement and magnitude of involvement.

For example, a capacity development initiative on land and water governance could include virtually everyone connected with irrigated agriculture (individuals, groups, formal organisations). Only very few need be involved in capacity assessments in the initial policy/concept development stages. There are a number of techniques, tools and methods that can support the stakeholder analysis such as surveys, workshops and conferences. The mechanism for stakeholder involvement is another key area. They may be represented through formal management/steering committees. Others may be represented through advisory or consultative councils/boards, surveys, workshops and conferences.

In relation to organisational capacity, the United Kingdom's Department for International Development has done a considerable amount of work and produced a useful document *Organisational sourcebook*. This suggests a number of different possible approaches to assessing organisational capacity, including the:

— open system model, which considers the strategic and institutional environment, the organisations inputs and resources, its strategy, culture, human resource management policies, systems, structure, outputs and performance;

— the 7-S model—strategy, structure, systems, shared values, skills, style and staff; and the

— organisational elements model covering inputs, processes, products, outputs, and outcomes.

It can be seen that all cover approximately the same ground, including most of the components itemised in the sublevels of the assessment matrix. All can be presented in the form of a table or matrix, making for easier visual understanding.

In the irrigation and drainage sector, specific aspects of organisational capacity may need to be considered, such as asset management policies and systems, quality management systems, financing mechanisms and accountability systems.

Assessing organisations within their institutional framework is of particular importance in the irrigation and drainage sector. This is because of the complex nature of irrigated agriculture and its linkage with so many other sectors of the economy and the social and cultural environment. A significant amount of work has been carried out in this regard by institutionalists such as Ms Ostrom. Working with others in areas of the humid tropics such as Sri Lanka, she has devised a set of 'institutional design principles' for irrigation organisations.

These design principles, which could also form the basis of assessing the capacity of existing organisations, are often liked by practitioners because of their relative simplicity and the clear-cut nature of the recommendations that flow from them. There is, however, a body of theory that is critical of the design principles, precisely because of their simplicity, suggesting that it is necessary to focus more on existing organisations and institutions and to take account of peoples' propensity to adapt existing institutions to address new and emerging issues (so-called 'institutional *bricolage*'). The complexity of the linkages between existing organisational mandates is a further factor to be taken into account, particularly in relation to the problem of the lack of fit between hydrological boundaries and other types, such as administrative boundaries.

Individual Level

Techniques for assessing capacity at the individual level are now well established. Indeed much of the criticism of capacity development programmes in the past is that they have been focused solely at the individual level, without paying sufficient attention to the organisational framework within which the individual operates.

Individual capacities are often assessed in relation to the need for training and lead to some form of assessment of training needs.

In the past, these were developed through a comparison of competencies and responsibilities and the need to try to match these two more closely for the organisation's human resources. Now there is a move to assess an individual's capabilities across a wider spectrum of criteria, such as attitudes and values, as well as the traditional focus on technical skills and competencies. These additional capabilities are summarised in the sub-levels of the individual capacity assessment matrix presented in Figure 1. There is increasing use of various types of psychometric testing, particularly in the corporate sector in industrialised nations, to assess capacities in this regard. However, as yet there are few references to the application of these techniques in the irrigation and drainage sector, particularly in developing countries.

General Assessment Tools

Certain assessment tools can be used at all three levels of capacity assessment. These include:

- Problem tree analysis, which is used to identify causes and effects of identified problems, and helps to look beyond visible symptoms to reach an understanding of underlying causes. The problem tree analysis may also be turned into an objective tree analysis, which forms the basis of a strategy to deal with the identified problems.
- Strengths, weaknesses, opportunities, threats (SWOT) analysis can be used by stakeholders to assess the enabling environment, the institutional framework and the specific organisation; and

— Benchmarking is increasingly used in the irrigation and drainage sector as a way of comparing performance and output over time, or with other organisations. It is a specific and structured method of performance assessment, which in its more general form can be used as a basis of assessing existing capacity without making comparisons with others or over time.

Strategic Options for Capacity Assessment

The process of capacity needs assessment is a step-by-step exercise, which first involves the assessment of existing capacity, then future capacity needs, followed by mapping capacity gaps, identifying opportunities for capacity development and defining objectives and targets as the basis for the formulation of a strategic plan of action. The proposed approach adopts a framework that is based on the principles of subsidiarity and participation. It analyses the capacity at three levels, of which the highest—the enabling environment—appears to be instrumental for a number of aspects including policy, legislation, socio-economic conditions and constraints to irrigation development. The middle layer addresses capacity needs of organisations involved in irrigation and the bottom layer aims to assess the needs of individuals.

The three levels provide a preliminary structure for capacity assessment and as an entry point for donors and technical assistance. Unfortunately, the framework remains somewhat general and idealised with little solid country experience that could support it, or assist in the development of a more sector-specific approach suited to the needs of a variety of systems, such as smallholder-based systems in Africa and large-scale canal-based irrigation in Asia.

Countries in sub-Saharan Africa and in Southeast Asia are probably among the most needy in terms of capacity assessment and development in agricultural water management. In sub-Saharan Africa, the irrigation sector is largely based on smallholdings of less than 1 ha, informally developed schemes, which are scattered over vast areas of land and around urban centres. Smallholders have no guaranteed access to water or to support services and training. Where still operational, official irrigation schemes are in need of rehabilitation and modernisation with not much hope that the investment costs will pay off.

Official government irrigation services and extension units have largely fallen into a state of dysfunction with no access to sufficient funds for transport or any irrigation related training materials. Government employees are subject to frozen salaries and/or stiff public entrenchment schemes. There is only little incentive and motivation to perform, as expected, these ambitious human resource development plans date back to the days of structural adjustment. Because of tight budgetary restrictions, these plans are now obsolete. This situation has created a substantial gap in public service provision to the sector. Partly the capacity gaps have been filled by national and international non-governmental organisations (NGOs) and an emerging private sector. However, NGOs often find themselves operating in a void being largely ignored by official policy-makers. Moreover, ambitious development projects are found that operate successfully over the project cycle. However, their impact on the capacity of their beneficiaries to manage irrigation systems is doubtful since long-term support of the beneficiaries is not guaranteed.

In several African countries, donors have encouraged governments to shift from state controlled to farmer-controlled irrigation management. No amount of process and organisational savvy will get a group of rural poor to accept a deal that does not make any economic sense to them. Important questions, concerning the viability of irrigated agriculture, must be addressed in conjunction with a change of management in irrigation.

Capacity assessment must be based on a broader conceptual framework, based on a deep appreciation of the prevailing irrigated farming systems. The conceptual framework must take into account the capacity for improved market integration, linkages with commercial institutions, access to credit, as well as improved on-farm water management practices.

Process Initiation

Before governments consider engaging in the process of capacity needs assessment there are a number of fundamental questions that need to be addressed and clarified. These include:

- Is capacity a binding constraint to the development of the irrigation sector?
- Does the sector have an irrigation strategy?
- Who drives the process (external versus self-assessment)?
- Is high-level commitment assured?

Is capacity a binding constraint to the development of the irrigation sector? This first, and probably most important, question concerns capacity as a binding constraint to the development of the sector. Is it binding, or are there constraints, other than capacity, that are as important or even more important to be addressed? Examples include access to markets, credit or affordable irrigation technology, etc. It is assumed that capacity assessment will not begin unless the need for it has been identified and subsequently been expressed. Ideally, such an expression of the need to initiate a capacity assessment process should be articulated by the irrigated farming community—the ultimate stakeholders. It should be supported through the strongest political commitment possible, reaching from the highest central levels to regional and district levels.

Does the sector have an irrigation strategy? The second question concerns the importance of irrigated agriculture in the context of the overall development of the country's irrigated sector, particularly in rural areas. Has government established an irrigation sector strategy that outlines the sector's long-term goals, defines clear objectives as to how the sector and its segments are to develop? Is the strategy well integrated into a country's water and agricultural sector policy and strategy? Such questions are of particular importance as many countries are now moving forward with the reform of the water sector taking an integrated water resource management approach. Donor-funded water sector reform initiatives are underway, which attract substantial funds. The irrigated sector, as the principal water user, is clearly a stakeholder in such initiatives and capacity assessment in irrigation must be streamlined with the broader planning in the water and agricultural sectors.

Who drives the process (external versus self-assessment)? Over many years, development agencies took the lead in the development of investment projects including capacity development components. Thereby critical steps such as the identification, formulation and design of projects were executed and driven by external experts, mostly consultants, assisted by national experts. The role and the involvement of developing countries were passive and largely reduced to that of recipients. Important opportunities for capacity development in project planning and management were wasted, since the process of preparing development projects or programme forms an important capacity development tool.

Increasingly donors move away from the concept of external capacity needs assessment towards the concept of self-assessment of capacity needs. The Global Environment Facility produced a guide for self-assessment of country capacity needs in global environmental management, in collaboration with multilateral donors including United Nations Development Programme (UNDP), United Nations Environment Programme (UNEP), the World Bank and FAO.

The National Global Environment Facility (GEF) Focal Point plays an important role in the initiation and facilitation of the self-assessment process. Focal points arrange meetings with high-level committees and with other national coordinating bodies to conduct preliminary discussions. Other actors may include high-ranking ministry policy decision-makers and representatives of key non-governmental organisations. The purpose of the initial self-assessment is to identify key capacity constraints, which can be used to determine interventions and activities to overcome these constraints (Figure 2).

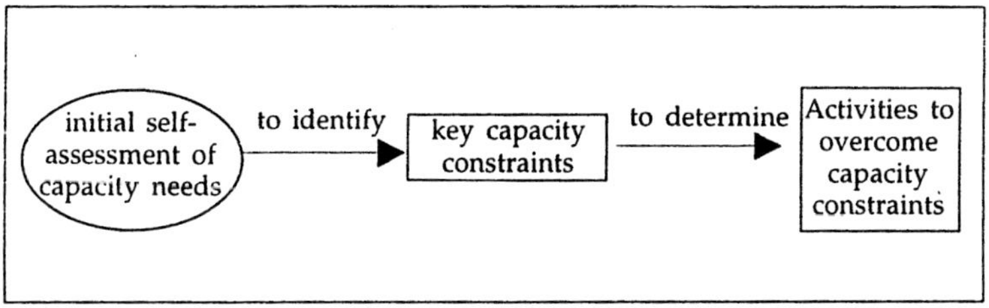

Figure 2. Global Environment Facility (GEF) approach to identify key capacity constraints and interventions

Assuming a government decides to give preference to a country-driven process, an important step would be to decide how the process can be successfully steered, effectively coordinated and efficiently administered, possibly with the assistance of an international facilitating body. Usually, the options include the setting up of a steering committee of high-ranking ministry and stakeholder representatives, a coordinating (apex) body, which can be established inside the leading ministry and with a clear mandate to drive the process, maintain the flow of information and communication with stakeholders.

In addition, there is a need for a facilitating stakeholder platform for stakeholders to express and articulate their interests and views precisely and without pressure. Preferably, such a platform would be established outside the ministerial structure in order to keep the balance of power between the disadvantaged and advantaged stakeholders. To drive the process successfully, a team of qualified individuals should be nominated, who are capable of organising and coordinating the process effectively and efficiently.

Is high-level commitment assured? Capacity assessment is not a static exercise. It involves a continuous process, which requires commitment to and 'ownership of change' on the part of high-ranking government officials, decision-makers and stakeholders as well as staying power and patience on the part of the actors. Commitment and ownership among senior management are regarded as essential, which can only be achieved if they are deeply involved and convinced that change is the ultimate alternative to stagnation. Moreover, commitment to change is a political statement that relies on the awareness, the active participation and the motivation of all stakeholders concerned.

Once fundamental questions are clear, and the capacity assessment process is initiated, important strategic decisions relate to the design of the process. This involves a number of critical decisions concerning clarity over the principles of stakeholder participation and involvement, and agreement on the objectives, a timeframe, proposed activities and mechanisms and finally the allocated budget.

Stakeholder Participation

There is a practical reason for promoting stakeholder participation in the process: to make more effective and efficient use of available and often limited resources. If stakeholders from the public, private sector, NGOs and civil society are involved, and adequately coordinated, it is likely that complementary qualities and synergies in experiences can be captured that otherwise may be overlooked.

Despite considerable stress on the involvement of non-governmental and private organisations in the process of capacity assessment, there appears to be little evidence that this is actually practiced, at least not in agricultural sector reform efforts Even where irrigation management responsibilities were transferred to water user organisations, government bodies appear to remain the principal beneficiary of capacity investments.

However, NGOs do offer important services to both formal and informal smallholder water management systems. If strengthened and integrated into an overall strategy, their impact on systems performance and food security could probably be enhanced and expanded to other areas. In addition, capacity in participatory irrigation project development and support, used by NGOs, could possibly be adopted as a model for more formalised extension and support services. As a consequence, NGOs and the private sector must be given due attention for capacity assessment in order to exploit best practices and approaches and identify their capacity needs more clearly.

Table 1. Categories of stakeholders

Category	Stakeholders
By level	National, intermediate (province, district) and grassroots level
By status	Public sector, private sector, non-governmental organisation (NGOs), public-private partnerships, community-based organisations (CBOs), water user associations (WUAs), technical assistance and donors
By location	Remote and isolated and non-remote (well connected) areas
By function	Clients, service providers, commercial, non-commercial and facilitators

The list of possible stakeholder and partners for capacity assessment in irrigation may differ from country to country depending on sector organisation and structure and on the administrative and institutional landscape. The overriding principle of participation implies that all stakeholders, who wish to be involved, get a fair chance to do so. However, too often a narrow sector view on irrigation results in little recognition of the fact that irrigation is part of a broader water resource management system. As many countries are now moving towards the concept of integrated water resource management, other water sector stakeholders must be included in the process.

In order to identify the stakeholders in irrigation it is useful to categorise stakeholders according to broad criteria and use these denominators to narrow the list of potential candidates involved in the process. The assumption is either their interests and positions are affected by poor performance of an organisation, e.g. irrigation services or that the organisation, individual or institution itself are in need of improved capacity. It is important to understand how irrigation systems are managed in which domain decisions are taken that can affect, or be affected, by stakeholders. Within the framework of irrigation, three broad areas may be identified:

— Water resource management domain, which may include the basin, sub-basins, command areas, farms, fields and even crops;

— Political administrative domain from the national to the local level;

— International domain, which cuts across the previously mentioned domains.

Within each domain stakeholders can be identified and their interests in and needs for capacity assessed. The following examples may illustrate the significance of this categorisation. At the national level, capacity is needed to prepare amendments for water laws, to establish policies on water tariffs and cost recovery mechanisms or to register water user associations. At the basin level, capacity is required to make water allocation decisions, or to impose restrictions on irrigation development within the basin. Similarly, the responsibility to administer irrigation schemes is usually vested at provincial levels and, therefore, officers need capacity to fulfil this function.

Within the domain of command areas, farmers need the capacity to make cropping and irrigation scheduling decisions, scheme operators need the capacity to provide water services and water user associations need management capacity in technical and financial matters. The distinction between administrative and water management domains is significant where boundaries of provinces or districts and command area do not match and an irrigation system falls under the responsibility of two or more administrative areas.

Almost all countries that share rivers with riparian neighbours are involved to some extent in regional cooperation and international initiatives on shared water resources. A country's capacity in irrigation planning and management is likely to be affected by such cooperation through the formulation of standards and directives. Effective cooperation may require in-country capacity in order to fulfil such requirements. Because of its crosscutting nature, all levels of the administrative and water management domains may be affected.

Other important stakeholders belonging to the international domain include unilateral and multi-lateral donors and international NGOs. Their interest in capacity development is justified through their desire to eliminate risks associated with the poor performance of irrigation projects. The analysis must further include the global environment that increasingly impinges on the capacity of a country from national to local level. An example is the way the Internet has created opportunities for information exchange; it also divides those with, from those without access.

Identification of stakeholders does not necessarily imply that all stakeholders are subject to rigorous assessment of current and future capacity needs. Probably, it will be necessary to limit the number involved to stakeholders from key domains including governmental, non-governmental organisations and the private sector. In a preliminary assessment, stakeholders may be categorised according to their possible future role in the process.

Options for Stakeholder Involvement

An important design decision concerns the stakeholders' future mode of involvement in the capacity assessment process. Options for stakeholder involvement can be categorised into three types:

1. *Advisory bodies*: Short-term body of national and international experts with a mandate to gather expert opinions on thematic issues. Such bodies may be composed of scientific boards and expert panels. The advantage of advisory bodies is the provision of informed views and recommendation on specific issues of mutual interest.
2. *Stakeholder inquiries*: Stakeholder inquiries intend to collect basic information on the institutional and organisational mandates, functions, performance and impact. Available tools include surveys, panel discussions and workshops. Stakeholder inquiries offer the opportunity to interact closely with the participants and address specific issues and concerns.
3. *Stakeholder forums*: This is a platform for a various group of stakeholders to engage in a facilitated consultation process, which permits open discussion and exchange of opinions, issues and concerns. The advantage of stakeholder consultation through such a forum is that it is arranged outside the usual ministerial bureaucracy. It may also offer a more conducive atmosphere for those stakeholders who are less confident and who feel intimidated by the presence of government representatives on the panel, even when the latter

are in a minority. Nonetheless, the success of public consultation platforms certainly depends on the quality of the facilitator who must be independent, communicative and articulate.

Identify Key Capacity Issues

An important step in the process of capacity needs assessment is to identify key capacity issues that need immediate attention and inclusion. Depending on the composition, structure and ability of stakeholders to articulate their perception of issues and problems, the list may be imprecisely formulated and long. However, being precise in the analysis of a problem is probably the most important step towards identifying priority issues.

The GEF guides for self-assessment of country capacity needs suggest a practical tool for the preliminary prioritisation of issues. In an issue prioritisation matrix, the identified and precisely formulated problems are listed and ranked against four criteria: the scale of the problem (local, provincial, national, international) its level of concern (low, medium high), the ability to adequately address the issue (low, medium and high) and finally priority ranking (one to five with one being most severe). For each priority issue, a focal problem needs to be identified and its root causes understood. Another step in the analysis is to clarify the effects and impact value systems have on the stakeholder.

Identifying Capacity Gaps

There seems to be some confusion concerning the assessment of existing and future capacity needs. The proposed approach takes the existing capacity at the level of individuals, organisations and the enabling environment as a reference (situation A) and compares it with future needs at the respective levels (situation B). The gap between the two situations A and B determines the capacity needs. However, the step from A to B poses a serious conceptual and methodological challenge and problem, especially in a multi-stakeholder and multi service-provider context.

The significance can best be described with an example: The identified issue shall be that smallholders in rural areas wish to adopt irrigation technology, but have no access to irrigation advisory and extension services. The causes of the problem may be identified as 'absence of official field staff, their restricted mobility and funding'. Another cause may be the inability of NGOs to expand to remote areas and to involve farmers there in their development approach. By applying the proposed approach, we would begin by assessing the existing capacity of each stakeholder. They would first be separately assessed and thereby their current constraints and limitations identified.

This would involve a number of questions concerning the organisation's mandate and its current performance in terms of effectiveness and efficiency. The next step would then involve the assessment of the future situation of each organisation. We would ask what the future mandate and vision shall be and what capacity is required to meet the future challenge. The comparison between the existing and future situation would then determine the capacity gaps in terms of resources and skills needed. Eventually, we would probably realise that no attention has been given to the needs of the farming community, the ultimate beneficiary or client.

However, the clients' needs determine the spectrum of services required to make smallholders' irrigation a viable enterprise. Equally important, it is assumed that service providers can offer the same range of services at comparable quality, which may not be the case. As a result, we may risk overestimating the capacity needs of each service provider and overlook synergies and comparative advantages.

Common sense suggests a different approach: To identify the needs of the clients, i.e. the irrigation farming community or the scheme operators first. Then assess the capacity of the identified service providers and their ability to meet the needs based on their assessed comparative advantages and capacity constraints. Once the services, rules and regulations needed are clarified, from the clients' perspective, the assessment of the service providers becomes more focused.

Taking a multi-stakeholder perspective, it is no more the difference between an organisation and individual's existing and future circumstances. Rather, a capacity gap is defined as the difference between the existing and future demand for a variety of services of multiple irrigation clients and the difference between the existing and future supply capacity of all service providers. The difference between existing capacity and futures capacity needs is referred to as the 'incremental capacity' and 'incremental service needs' respectively. Both are related to each other and determine the capacity gap.

In the context of a multi-cultural society, it is appropriate to begin the capacity assessment with the local and indigenous knowledge of an organisation and its members and enhance its understanding and appreciation. The concept of developing existing capacity rather than importing it is conventionally known as 'organic capacity development'. It is based on sharing knowledge and coaching.

Capacity Opportunities

Assessment of the future situation becomes more complicated when considering that the future performance of an organisation and/or individual is partly influenced by the external environment. Exposure to the outside world can create opportunities for the organisation to enhance its capacity to take an opportunity onboard. For example, it is now well established that with the arrival of the Internet, organisations involved in irrigation research and extension have the opportunity to access relevant information and to communicate more easily with networks of experts and colleagues.

Developing advanced tools, such as remote sensing data and tools and geographic information systems (GIS), and the capability to apply them in the management of water resources will offer new opportunities to resource managers. Thus, they become more responsive to the needs and aspirations of the user communities they service. The adoption of the information tools is challenging conventional notions of how government and non-governmental agencies should work and relate to each other.

Capacity Needs of Complex Water Service Systems

Larger irrigation and drainage systems are complex socio-technical water service systems with a full array of different stakeholders, institutions and players. Ideally, the purpose of service provision is to meet the water service needs of the irrigated crop production system and keep the farmers—the ultimate clients—happy. To do so, service providers need to have the capacity to provide the services as demanded and to the client's satisfaction.

Key services, roles or functions directly involved in irrigation management service provision could be categorised into primary, secondary and other services. Primary services include the provision of the hydraulic infrastructure, the water delivery service (water capture, water conveyance, water distribution, water allocation etc.). Secondary services comprise those that are integral to the primary services such as operation and maintenance), supporting services (information provision, coordination, representation, etc.). Other services include the arrangement for primary, secondary and supporting services (i.e. to select providers, to define terms of reference, conclude agreements or contracts, monitor provisions etc.).

Relations between service providers and clients are not always clear. Many players and actors are involved who make transactions without formal recognition of their importance in the overall performance of the irrigation system. Huppert *et al.* provide a comprehensive set of tools for analysing the service relations in the provision of maintenance services in irrigation and drainage. An example for a visualised analysis of the institutional arrangements for service provision is given in Figure 3.

According to their conceptual framework, the provision of any irrigation service involves interactions between service providers, service recipients, service payers, government policy-makers and regulators. Typical irrigation services include adjusting gates, measuring and distributing water, or maintaining canals and hydraulic facilities. In return, water users (clients) make compensations in the form of cash payments and in-kind transfers. The capacity of the service provider to fulfil the client's needs depends on their ability to interact with the client effectively and to respond to the needs in a professional, flexible and reliable manner.

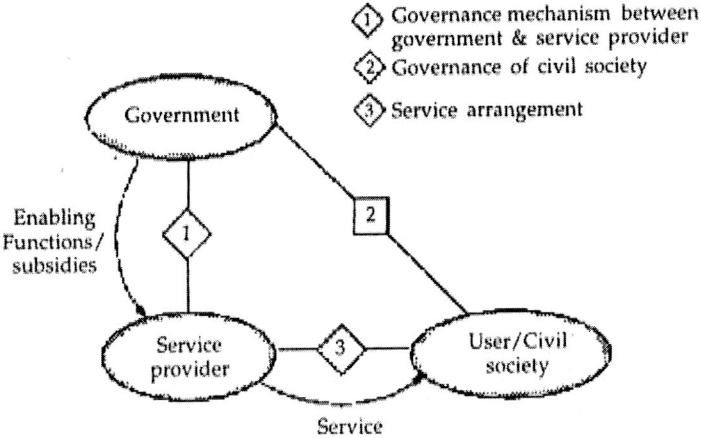

Figure 3. Institutional arrangements for service provision

Briefly, the assessment of capacity needs of service organisations in irrigation must devote special attention to both the 'provider system' as a conglomerate of many players and to the 'client system', which is heterogeneous and divers. The capacity to communicate effectively plays an important role in order to make irrigation service provision a success.

Clarity of Purpose and Impact

A critical design condition for the design of capacity needs assessment in agricultural water management is the need for a logical framework. Only if an identified lack of capacity can be clearly linked to a well-defined purpose such as a strategic sector goal, an objective or a development target, can the full support of stakeholders and donors be assumed. For example, if the capacity development objective for a comprehensive capacity assessment of the official irrigation service system is to improve the quality of service provision, the purpose should be improved client satisfaction and eventually improved productivity of water use. A positive impact of improved capacity in irrigation service provision on client satisfaction and increased water productivity must be supported by a logical framework approach in which needs are linked to a possible increase in suitable performance indicators.

Clarity of impact is a concept that requires the anticipation of possible changes that the improved capacity would have on the performance of individuals, organisations and on society. This is to be linked to the economic, social and environmental changes of individuals and organisations. Indicators are required at all levels to be able to quantify and demonstrate that the identified capacity gap is the root cause for the poor performance at the level of the ultimate beneficiaries (target group level). They are the ones who are intended to benefit from the improved performance, resources and framework condition.

A worked example of a logical framework is given by the International Fund for Agricultural Development (IFAD) in which specific needs of capacity development are taken into account. It is explained that capacity intervention require a different set of performance questions, target indicators and monitoring mechanisms as compared to infrastructure development projects.

Managing Capacity Assessment

When looking at the geographical map of many countries, it becomes apparent that actually irrigated and potentially irrigable areas are scattered over vast land areas, with long distances between and insufficient infrastructure in place. Obviously, not all areas can be included in capacity assessment and development programmes. A full coverage of remote areas would simply reach beyond the scope and resources available for capacity assessment initiatives. It is therefore obligatory to match the scope of the assessment with available time, qualified staff and funds. A restriction of areas implies that to establish and agree on criteria that exclude certain areas and domains from the exercise, the following options may be taken into consideration: high potential versus low potential areas, isolated versus well connected areas.

By restricting the area covered by the capacity assessment there is a substantial risk that the important role of irrigation and agricultural extension field staff in the development of rural areas may be overlooked. Capacity assessment of field staff may be crucial where mobilising and facilitation of communities is crucial for the successful adoption of grassroots programmes such as farmer field schools.

Dealing with Insufficient Incentives

It is now widely recognised that insufficient payment of civil servants and insufficient funding for operations and service provision are the most significant constraints to high performance and job satisfaction. Very often highly skilled and educated individuals in the public sector find themselves trapped in a system that barely pays for subsistence needs. Funds for operations and day-to-day works have dried up and there is often little hope for a change for the better. As a result, the time at work is used for non-professional business, or it is used unproductively.

Donors tend to offer incentives in the form of salary supplements and travel allowances, which may result in a drain of the qualified ministry personnel, create inequity in pay, jealousy and may further reduce the motivation of those who are left behind.

The implication of a loss of motivation has a direct influence on the performance of an organisation. The latter is determined by its internal capacity in conjunction with its internal motivation. Performance is also determined by external factors such as the legal, social and economic context. According to Horton, organisational capacity refers to the staff complement or human resource capital, as well as to its structure, management system and linkages with other organisations. Organisational motivation refers to the organisational cultures and incentives that influence the use of capacities in pursuit of the organisation's goals. The principal relations between performance, internal capacity and motivation and external effects of and on the environment are given in Figure 4.

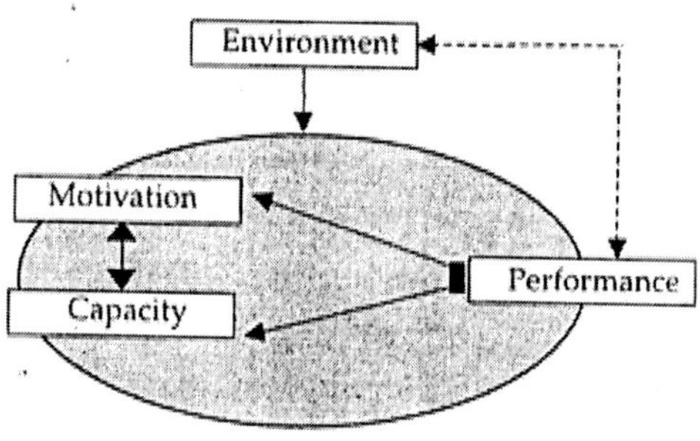

Figure 4. Relationship between an organisation's environment, motivation, capacity and performance

The implication for capacity assessment is that it must not overlook the importance of the organisation's motivation and incentive system for effective and efficient job fulfilment of its members. In a comprehensive assessment, the human development system must be made subject to a rigorous review and reform. Otherwise, it cannot be assumed that identified capacity needs are the root cause of bad performance. Lopes and Theisohn suggest there is need for a locally led, harmonised reform of the way public incentives are structured and integrated into the national budget.

An example for identified incentive deficiencies in irrigation and drainage maintenance service provision adopted from Huppert *et al.* shows they play an important role in the way irrigation and drainage system maintenance is performed. The deficiencies cut across the entire spectrum of stakeholder from senior irrigation officials, who regard benefits of maintenance as too little, to water users who see maintenance as a responsibility of the owner of the property, which, in this case, is the government.

Managing Resistance to Change

Resistance to change is widely recognised as a sector-crosscutting problem in the domain of development and, in particular, in capacity development. Within complex organisations such as ministries and irrigation agencies, individuals and lobby groups can always be found that actively oppose change and pursue a hidden agenda. Such behaviour is human and as such recognised as intangible.

Vested interest and power differentials may have the greatest influence on the progress made in capacity development. The mindset and the personalities, informed by language and culture; dictate the course of communication and collaboration. They create virtual realities that influence how people, institutions and societies behave. In addition, they indicate that any intervention involving a transfer of resources is potentially subject to numerous influences and corruption. The strategies to manage resistance proposed by DFID entail a number of measures at different levels. Grouped into three categories, each requires different activities: 'unwilling' (activities will include negotiation, persuasion, introduction of positive and negative sanctions); 'unable' (this will require training, skills transfer, support and coaching); and 'unknowing' (activities will focus on communication and involvement).

Moreover, probably one of the most important strategies in capacity needs assessment is attaining transparency of the process and accountability of the actors involved. Transparency implies clarity of the process at each step taken, close communication with all stakeholders and effective facilitation of the process.

Effective facilitation is a broad range of knowledge, attitudes and skills in working with diverse groups in development settings. These skills provide the basis for more effective interventions to strengthen each stage of the activity cycle. Good facilitation work builds ownership and commitment to results and increases the likelihood that activities will respond to participants' actual needs. A good communications plan should consider the audience, message, media, frequency, goals, responsibility, feedback mechanisms and measures of communication effectiveness.

7

Drainage Water Management

Large-scale development of irrigation has taken place in many arid and semi-arid areas since the late nineteenth century. Although irrigation has greatly increased the agricultural production potential, recharge brought about by seepage losses from the irrigation network and deep percolation from farm irrigation has accumulated into the underlying groundwater. A rise in water table results when irrigation-induced recharge is greater than the natural discharge.

In many irrigated areas around the world, rising water tables have subsequently led to waterlogging and associated salinity problems. This has happened where drainage development has not kept pace with irrigation development or where maintenance of drainage facilities has largely been neglected. Salinisation affects about 20-30 million ha of the world's 260 million ha of irrigated land. To maintain favourable moisture conditions for optimal crop growth and to control soil salinity, drainage development is indispensable especially in saline groundwater zones.

Current drainage improvement programmes cover less than 0.5 million ha per year, insufficient in their view to balance the current growth of affected drainage areas.

They estimate that: 10-20 percent of the irrigated land is already equipped with drainage; 20-40 percent of the irrigated area is not in need of any artificial drainage; while 40-60 percent is in need of drainage but remains without drainage facilities.

In Central Asia, the present drainage infrastructure is insufficient to control irrigation-induced waterlogging and salinity with a comparatively small percentage of subsurface drained land. In addition, the poor state of drainage networks (due to lack of maintenance) has exacerbated waterlogging and salinity.

In the Near East, which is a region subject to salinity problems due to the prevailing climate conditions, an average of about 29 percent of the irrigated areas have salinity problems. 12 countries in the Near East on average about 34 percent of the irrigated area has been provided with drainage facilities. For most countries no figures are available on the area under surface versus subsurface drainage.

In Pakistan, 13 percent of the irrigated area is reportedly suffering from severe salinity problems in spite of the efforts made to provide drainage in irrigated areas. Salinity problems persist because of deficiencies in water policies and the low priority attached to the allocation of resources for the operation and maintenance (O&M) of drainage facilities in favour of initiating new projects.

On the other hand, under the influence of the growing world population and the increasing demand for food, there is a trend of irrigation intensification. To supplement scarce surface water resources, groundwater is exploited through tubewell development, mainly in fresh groundwater zones, all over the world. In many of these areas, the water table is declining due to overexploitation of groundwater resources.

Therefore, problems of waterlogging and related salinisation in irrigated agriculture are confined principally to saline groundwater zones. However, the salinisation and sodification of agricultural lands resulting from irrigation with marginal and poor-quality water is increasing rapidly. Although firm figures are not currently available, many cases have been reported and documented for major irrigated areas in the world including South Asia, Southeast Asia, Central Asia, North Africa, the Near East, Australia, and the United States of America.

Importance of Water Conservation and Reuse

In response to the increasing world population and economic growth, water withdrawals for human consumption will increase, so increasing the competition for water between municipal, industrial, agricultural, environmental and recreational needs. If present trends continue with water withdrawal under present practices and policies, it is estimated that by 2025 water stress will increase in more than 60 percent of the world.

In this respect, providing food for the growing population is a major challenge as agriculture is already by far the largest water consumer in most regions in the world, except North America and Europe. On a global basis, agriculture accounts for 69 percent of all water withdrawals. Although the water resources for agriculture are often overused and misused, the general belief is that irrigated agriculture has to expand by 20-30 percent in area by 2025 in order to produce sufficient food for the growing world population. In order to avoid exacerbating the water crisis and to prevent considerable food shortages, the productivity of water use needs to increase. In other words, the amount of food produced with the same amount of water needs to increase. This is possible through the conservation and reuse of the available water resources in the agriculture sector, including usable drainage waters.

The overuse and misuse of water in irrigated agriculture has not only resulted in large-scale waterlogging and salinity and overexploitation of groundwater resources, but also in the depriving of downstream users of sufficient water and in the pollution of fresh water resources with contaminated irrigation return flow and deep percolation losses. Water pollution adds to the competition for scarce water resources as it makes them less suitable for other potential beneficial downstream uses. Furthermore, it might cause severe environmental pollution and threaten public health

Until ten years ago drainage water management received little attention. Drainage research tended to focus on design issues, while evaluation dealt largely with the performance of the installed system in relation to the design criteria. After the 1992 Earth Summit, the international irrigation and drainage community focused its full attention on drainage water management.Agenda 21 not only stresses the need for drainage as a necessary complement to irrigation development in arid and semi-arid areas, but at the same time it urges the conservation and recycling of freshwater resources in a context of integrated resource management. In many countries these concerns and especially the concern for water quality degradation have resulted in drainage water disposal regulations to maintain the water quality standards of freshwater bodies for other uses, i.e. agricultural, industrial, environmental and recreational uses.

System Approach in Drainage Water Management

When planners, decision-makers or engineers face the need for a change in drainage water management, the nature of the exact problem is often not clear and the perceived problem depends considerably on the individual's viewpoint.

For example, a farmer's perception of a problem related to drainage water management depends on the physical conditions within the farm boundaries and may differ substantially from those of an irrigation district, national water resource authority or environmental pressure group. In such situations, a soft system approach can help define the problem and seek solutions. A key feature of the soft system approach is that it attempts to avoid identifying problems and seeking solutions from only one perspective and excluding others.

The characteristic of a system is that, although it can be divided into subsystems, it functions as a whole to achieve its objectives. The successful functioning of a system depends on how well it satisfies changing external and internal demands. The system itself is part of a broader universe. In all natural or agricultural systems, there exists a hierarchy of levels. For irrigated crop production this hierarchy might be:

— biochemical and physical systems;

— plant and cropping system;

— farming system;

— irrigation and drainage systems;

— regional, river- or drainage-basin system; and

— supra-regional systems.

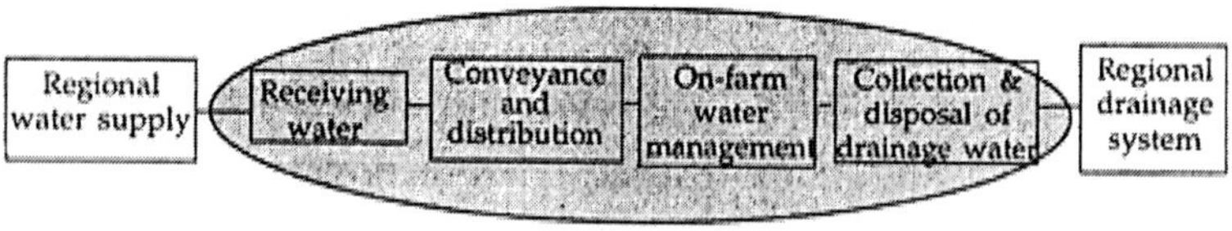

Figure 1. Example of a water management system within an irrigation district

Figure 1 provides a simplified example of a water management system within an irrigation district. It shows some main characteristics of a system as described by the Open University.

1. Defining the boundaries of a system is not a simple task. For example, there could be questions as to whether it would be better to include the regional drainage system within the system boundaries and include pressure from environmental groups, other water users and water quality rules and regulations in the system's environment. The boundaries chosen reflect the perception of the systems analysts and their understanding of the system's behaviour, which may not always coincide with an organisational or departmental unit. System boundaries in drainage water management might more often coincide with hydrological units.
2. The elements included in the system's environment are those which influence the system in an important way but over which the system itself has no control because they are driven by forces external to the system of interest.
3. The elements within the system are functional working parts of the system. The way that the system operates and behaves depends on the interactions between the elements in the system. Elements could be decomposed into smaller

subsystems. The level of detail depends on the specific objectives of the systems analysis. For example, in systems analysis for drainage water management, it would be important to show individual farms as subsystems because on-farm water management defines to a great extent the total quantity and quality of the generated drainage water.

4. As the creator of the system image, a systems analyst defines a particular perspective while another analyst of differing disciplinary training may produce a different image.

When applying a systems approach, at first it is necessary to include all elements in the picture whether they relate to physical, technical, economic, legal, political or administrative considerations as well as any subjective considerations based on the understanding, norms, values and beliefs of the stakeholders involved. A premature exclusion of important elements may result in a suboptimal course of action.

Where changes in drainage water management are necessary, a simple method to produce predetermined results might not work. Rather, a more open-ended investigative approach is required in which important avenues are explored and considered and in which there is room for iterative processes. The outcome of such an investigative approach is not necessarily the optimal final solution but rather a solution that seems best for those involved, for that particular time and under those particular circumstances. As the external environment and internal demands change constantly, the mix of solutions and needs for change are changing continuously. Moreover, the achieved changes themselves give new insights into processes and enable a continuous learning process.

The soft system approach to identifying causes and seeking solutions takes the aforementioned considerations into account. The soft system methodology is a seven-stage approach that has been adopted and adjusted for numerous purposes. The soft system methodology was developed for managing changes in the context of human activity systems.

Problems in Drainage Water Management

Defining the problem and diagnosing the causes is the key to seeking solutions. In the context of drainage water management a problem might be defined as: a need, request or desire for a change in present situation subject to a number of conditions or criteria that must be satisfied simultaneously. This definition implies that a problem situation does not exist independent of the stakeholders who perceive the problem. The solutions to the problem are subject to criteria and conditions based on people's perception of the ideal situation. The conditions and criteria are subjective insofar as they are based on personal or communal objectives influenced by local to national economic, social, cultural, ecological and legal motives, norms and values.

In the field of agricultural drainage water management, the root cause of the problem(s) nearly always stems from human interference in the natural environment. An analysis of the problem and the search for solutions starts in a situation where someone or a group of people perceives that there is a problem. At this stage, it might not be possible to define the problem with precision, as different people involved will have differing perceptions of the problem.

The first step in formulating the problem more precisely is to identify all the stakeholders involved and their relation to the problem. To form what is called the rich picture, all the elements must be included whether they relate to physical, technical, economic, legal, political or administrative considerations along with subjective considerations based on understanding, norms, values and beliefs of the stakeholders involved. It is then necessary to extract areas of conflict or disagreement as well as the key tasks that must be undertaken within the problem situation.

Once the problem situation has been analysed and expressed from the key issues, the relevant systems and subsystems can be defined. These systems can be formal or informal and are those that carry out purposeful activities that will lead to improvement or elimination of the problem situation. Examples of a relevant system within the context of drainage water management might be an on-farm water management system, a system for development of water quality rules and regulation for drainage water disposal, or a land-use planning system. An analysis of the various elements involved provides valuable insight into different perspectives on and constraints surrounding the situation. For each relevant system, a root definition can be formulated.

A root definition is a formulation of the relevant system and the purpose of the system to achieve a situation in which the problem is balanced out or eliminated. Each root definition provides a particular perspective of the system under investigation. In general, a root definition should include the following information: what the purposeful activity carried out by the system is; who the 'owner' of the system is; who the beneficiaries/victims of the purposeful activity are; who will implement the activity; and what the constraints in its environment are that surround the system. A root definition for on-farm water management might be:

> In an on-farm water management system the responsible farmer uses irrigation water in such a manner that the drainage water generated is of such quantity and quality as permitted by the drainage disposal act whilst maintaining long-term favourable soil conditions that guarantee the production of valuable crops to ensure the financial sustainability of the farming enterprise.

A root definition for the development of water quality rules and regulation could be:

> A governmental system in which water quality rules and regulations for the disposal of agricultural drainage water are promulgated such that they will guarantee water quality for beneficial downstream water uses, including the maintenance of valuable ecosystems, while ensuring the economic sustainability of the agriculture sector.

Seeking Solutions

On the basis of the root definitions, conceptual models need to be constructed. These models include all the probable activities and measures that the system needs to implement to achieve the root definition. In other words, alternative scenarios need to be formulated. This involves doing sufficient work on the technical and other details, which need to be defined, in order to enable sound decision-making. Moreover, measurable indicators need to be established to compare the results of the conceptual models with the analysed situation or base case. As the formulation of scenarios is based on a thorough analysis of the systems, it should take into consideration system objectives, possibilities and constraints.

Improvements needed to minimise or correct a particular drainage water management problem may consist of physical structures, non-physical improvements or both. Physical improvements could involve using irrigation water conservatively by on-farm water management practices along with regional drainage practices such as recirculating usable drainage water to meet waste discharge requirements. Non-physical improvements may include implementing tiered water pricing to encourage growers to use water wisely, i.e. charging a penalty for overuse.

The next step is to compare the scenarios or conceptual models with the situation analysis. The idea is to test the scenarios and decide whether the implementation of a scenario would resolve the defined key issues. If it would, it needs to be investigated and there needs to be debate as to whether the changes proposed, resulting from implementation of the scenario, are both desirable and feasible. What is desirable and what is feasible might clash as a result of system objectives, possibilities and constraints. The final step is to define the measures and changes to be implemented.

Spatial Issues

Drainage water problems vary in space and time due mainly to soil heterogeneity and water management practices. The following example illustrates how spatial variability needs to be taken into account in the problem analysis, and also how it influences the options for drainage water management.

Environmental problems related to agricultural drainage water disposal on the western side of the San Joaquin Valley, California, the United States of America, have created a need for improved irrigation water and salt management. The presence of harmful trace elements, mainly selenium, in the drainage water is of major concern and has led to limitations on drainage water disposal to rivers and impoundments such as agricultural evaporation basins. Where water districts in the problem area fail to meet selenium and salt load targets, they risk monetary penalties and loss of access to disposal sites.

Young and Wallender raised the question of whether the constraints raised by current or future regulations will reduce drainage to the point where salt accumulation would occur and in which areas this might occur first. Second, they raised the question of the spatial distribution of drainage water disposal costing strategies. To answer these questions, they developed a methodology for the Panoche Water District to calculate the spatial distribution of water-, salt- and selenium-balance components using data collected by the water district. Furthermore, they developed and evaluated spatially distributed drainage water disposal costing strategies. The following is a brief overview of their research findings.

The Panoche Water District, situated in the western side of the San Joaquin Valley, is a typical example of a water district that needs to cope with disposal limitation in the form of selenium load targets. The groundwater in the Little Panoche Creek alluvial fan contains sodium-chloride type water, relatively low in salt content, and with selenium concentrations ranging from 1 to 27 µg/litre. In contrast, the groundwater in the Panoche Creek alluvial fan contains sodium-sulphate type water, relatively high in salt content, and with selenium concentrations ranging from 20 to 400 µg/litre. Due to overirrigation since the introduction of surface water delivery systems in the 1950s, the water table rose to within 1-3 m of the ground surface.

Subsurface drainage was installed which maintained successfully the water table at an acceptable level for agriculture. However, due to irrigation and drainage practices the naturally occurring salts and selenium in the region's soils are mobilised and enter into the subsurface drainage system as well as into the shallow groundwater.

Calculation of a spatially distributed water balance revealed that downslope areas with shallow water tables receive groundwater discharge. Drains in these areas intercept lateral and vertical upward flowing groundwater, while the upslope undrained areas recharge to the groundwater.

The highest salt load in the collected drainage water occurred in the centre and northwestern part of the district that corresponds with the location of greatest drainage. Salts entered the drainage water via the groundwater with the maximum occurring at the alluvial fan boundaries. Accumulation of salinity occurred largely in the drained regions with the maximum occurring roughly in the regions of maximum groundwater recharge. In the undrained regions, more salt was removed from storage as compared to the drained regions, caused by greater deep percolation in the undrained areas coupled with salt pickup.

The amount of selenium removed by the drains was greatest on the Panoche Creek alluvial fan and the interfan. The drainage system removed selenium through groundwater discharge. Selenium storage in the undrained areas decreased in proportion to the volume of deep percolation, while in drained areas selenium accumulated in areas similar to those where total salt storage increased.

Assuming a charge on drainage volume, spatial distribution of a drainage penalty per hectare would unfairly affect growers on the Little Panoche alluvial fan where relatively little selenium originates. In contrast, a charge levied per kilogram of selenium discharged from the drainage systems would result in growers on the Panoche Creek alluvial fan and the interfan paying more for drainage disposal in accordance with the higher selenium loads. Neither of the two methods of assessing drainage penalties addresses the poor water management in the upslope undrained areas that contribute to downslope drainage problems. A more equitable charge on the amount of selenium discharged into the environment from a control volume would account for excessive deep percolation as well as reflect differences in selenium loading caused by geological variations.

Use of Models in Recommending Solutions

A major distinction is often made between simple and complicated models in which the former is frequently associated with engineering methods and the latter with scientific methods. The development of these different types of models and the use of the terms stem from the needs of various groups of professionals. Engineers, managers and decision-makers are in general looking for answers and criteria to base their management, decisions or designs on, while scientists are more interested in the underlying processes. The terms simple and complicated in relation to engineering or functional and scientific models are rather subjective. The distinction between scientific and functional refers not only to the purpose of modelling and the intended uses, but also implicitly to the approaches on which the models are based.

The three main groups of modelling approaches are mechanistic, empirical and conceptual approaches. Mechanistic, or as Woolhiser and Brakensiek define them, physically-based models are based on known fundamental physical processes and elementary laws. In groundwater modelling, this approach is also known as the Darcian approach. As this approach is based on elementary laws it should be, theoretically, valid under any given condition and therefore its transferability is extremely high.

On the other hand, empirical approaches are based on relations that are established on an experimental basis and are normally only valid for the conditions under which they have been derived. Finally, conceptual calculation approaches are based on the modeller's understanding of fundamental physical processes and elementary laws, but these are not used as such to solve a problem. Instead, a concept of the reality is used to tackle a problem. The best-known example is the bucket-type approach to describe the flow of water through unsaturated soil.

Scientific models make use of mechanistic calculation approaches whenever possible and avoid the use of empirical and conceptual approaches. In contrast, functional models might include any of the three calculation approaches. Here, mechanistic approaches might be included as long as they do not conflict with other required model characteristics such as simplicity and short calculation time. Empirical and conceptual approaches are used in functional models as the only concern is that the model serves its intended purpose.

Regional models

The employment of a range of drainage water management options results in certain benefits, interactions and trade-offs not only in the place where the measure is implemented but also in adjacent and downstream areas. To enable decision-makers, managers and engineers to choose from different options and to study the effects of various alternatives, regional simulation models will need to be employed. Regional models normally include three main calculation modules, i.e. water flow and salt transport in the unsaturated or vadose zone, through the groundwater zone and through the irrigation and drainage conduits. Regional models normally require large amounts of data, and model calibration and validation is a time-consuming exercise.

Rootzone hydrosalinity models

Rootzone hydrosalinity models may range from simple conceptual to complex scientific models. In the more simple models, the spatial component in the control volume is typically assumed to be homogenous and space averaged (lumped), but water flow pathways are treated as distributed fluxes. The time increments taken may vary from irrigation intervals to crop growing season. Salinity is often treated as a conservative (non-reactive) parameter in simple models. The advantages of simpler functional models include more limited requirements for input data and model coefficients.

In contrast, the more complex, process-based rootzone models simulate water flow based on Richards' equation and treat salinity as a reactive state variable with simplified to comprehensive soil chemistry submodels. Such models provide greater understanding of the complexities in interactive physical and chemical processes. Complex scientific models require extensive input data and model coefficients, and carry out computations over small time and spatial scales.

A word statement of Richards' equation for the rootzone may be given by:

[Rate of change in soil water with respect to time] = [Rate of change in flux with respect to depth] - [Root water extraction sink with respect to depth and time]

In one dimension and taking small soil volume elements, Richards' equation is:

$$\frac{\partial \theta}{\partial t} = \frac{\partial}{\partial z}\left[K(\theta)\frac{\partial H}{\partial z}\right] - S_e(z, t) \quad (1)$$

The terms in the parenthesis represent flux taken as the product of hydraulic conductivity (*K*) and hydraulic head gradient (¶*H*/¶z). Richards' equation is difficult to solve because there are two dependent variables volumetric water content (*q*) and *H*, the relationship is non-linear as *K* is a function of (*q*), and the water extraction sink (S_e) requires simulation of root growth by soil depth (z) and time (t). Once the soil water flow is simulated, the output data (*q*, flux) serves as input data for simulating soil chemistry.

The reactivity and transport of chemical species is obtained from:

$$\frac{\partial \theta C_k}{\partial t} + \rho\frac{\partial \overline{C}_k}{\partial t} + \rho\frac{\partial \overline{C}_k}{\partial t} = \frac{\partial}{\partial Z}\left[\theta D\frac{\partial C_k}{\partial Z} - qC_k\right] \quad (2)$$

k = 1, 2,..., n

The principal solute species (C_k= 1..n) modelled are sodium (Na), calcium (Ca), magnesium (Mg), potassium (K), chloride (Cl), sulphate (SO_4), bicarbonate (HCO_3), carbonate (CO_3) and nitrate (NO_3). r is bulk density, D is dispersion coefficient, q is soil water flux, is the exchangeable form and is the mineral form of the solute species.

An early hydrosalinity simulation model was later extended to the widely used LEACHM. The US Salinity Laboratory has been active in modelling efforts for salt transport and major cations and anions such as UNSATCHEM and HYDRUS, both in one and two dimensions. Trace elements of concern such as boron and selenium have yet to be incorporated into these models. Simultaneous water, solute and heat transport modelling of the soil-atmosphere-plant continuum has been developed at Wageningen Agricultural University, the Netherlands, in collaboration with ALTERRA (formerly the DLO Winand Staring Centre). The present version, SWAP 2.0, integrates water flow, solute transport and crop growth according to current modelling concepts and simulation techniques.

Salt balance in the rootzone

The long-term sustainability of irrigated agriculture is heavily dependent on maintaining an adequate salt balance in the crop rootzone. For regions with a high water table, the salt balance needs to be expanded to include the shallow groundwater, too. Kaddah and Rhoades examined salt balance in the rootzone subject to high water table with:

$$V_i * C_i + V_{gw} * C_{gw} + M_f =$$

$$V_{dw} * C_{dw} + M_p + M_c + \Delta M_{ss} + \Delta M_{xc} \quad (3)$$

where:

V_i = volume of irrigation water (m^3);

V_{gw} = volume of groundwater (m^3);

V_{dw} = volume of drainage water (m^3);

C_i = salt concentration of irrigation water (kg m^{-3});

C_{gw} = salt concentration of groundwater (kg m^{-3});

C_{dw} = salt concentration of drainage water (kg m^{-3});

M_d = mass of salts dissolved from mineral weathering (kg);

M_f = mass of salts derived from fertilizers and amendments (kg);

M_p = mass of salts precipitated in soils (kg);

M_c = mass of salts removed by harvested crops (kg);

ΔM_{ss} = mass of changes in storage of soluble soil salts (kg); and

ΔM_{xc} = mass of changes in storage of exchangeable cations (kg).

Cation exchange between calcium, magnesium and sodium may modify the balance of these cations in the soil water and affect mineral solubility. Sodium minerals are more soluble than calcium minerals while magnesium minerals may range from highly soluble (sulphate type) to sparingly soluble (carbonate type).

Equation 3 contains some components that are not known or are small in relation to other quantities such as M_d, M_f, M_p and M_c. Moreover, the sources M_d and M_f tend to cancel the sinks M_p and M_c. If steady-state conditions are assumed for waterlogged soils, ΔM_{ss} and ΔM_{xc} may be assumed to be zero so that Equation 3 reduces to:

$$V_i * C_i + V_{gw} * C_{gw} = V_{dw} * C_{dw} \tag{4}$$

If the land is not waterlogged, $V_{gw} * C_{gw}$ drops out so that salt balance can be viewed simply and leads to such relationships as:

$$LF = \frac{V_{dw}}{V_i} = \frac{C_i}{C_{dw}} \tag{5}$$

Equation 5 expresses the leaching fraction (*LF*) and is the simplest form of the salt and water balance for the rootzone where surface runoff is ignored. Where the land is not waterlogged, V_{dw} consists of deep percolation from the rootzone.

Where the water table in agricultural lands is controlled by subsurface drainage, then the mass of salts in groundwater must be considered in Equation 4. Due to the nature of flow lines to subsurface drainage collector lines, the subsurface drainage collected and discharged is a mix of deep percolation from the rootzone and intercepted shallow groundwater. For example, for the Imperial Irrigation District, California, the United States of America, Kaddah and Rhoades estimated that deep percolation contributed 61 percent and shallow groundwater 39 percent to the tile drainage effluent based on chloride mass balance. They also estimated that tailwater contributed 10 percent to the total surface and subsurface drainage from the district. The ratio of tailwater plus intercepted deep percolation and shallow groundwater to applied water for the district was 0.36.

In the presence of high water table, shallow groundwater and its salts may move up into the rootzone (recharge) and down out of the rootzone (discharge) depending on the hydraulic head. Deficit irrigation under high water table may induce rootwater extraction of the shallow groundwater. The salinity level of the shallow groundwater is of some concern under such conditions. However, there does not appear to be a simple conceptual model of capillary rise of water and solutes.

Various versions of the salt balance, Equation 3, have served as the basis for numerous models including SALTMOD, CIRF and SAHYSMOD, which is under preparation by the International Institute for Land Reclamation and Improvement, Wageningen, the Netherlands.

8

Economic Evaluation of Agricultural Water Resources

Water has unique characteristics that determine both its allocation and use as a resource by agriculture. Agricultural use of water for irrigation is itself contingent on land resources. In a situation of growing water scarcity and rising demands for non-agricultural use of water, reassessment of sectoral allocations of water are inevitable. In developing countries, irrigated agriculture plays a vital role in contributing towards domestic food security and poverty alleviation. Therefore, achievement of these objectives is dependent on adequate allocations of water to agriculture.

Justification of such allocations requires that irrigated agriculture be a cost-effective means of achieving stated political or social objectives, such as food security or poverty alleviation, and that all externalities be taken into account in the pricing mechanism. Improved allocation of irrigation water is required within the agriculture sectors of developing countries in order to achieve greater efficiency in the use of irrigation water and existing irrigation infrastructure. Reallocation is also required in order to reduce waterlogging and salinisation of irrigated land, to decrease the negative environmental impacts and other externalities of irrigation. Water provides goods and services that are utilised by agriculture, industry and households. Provision of many of these goods and services is interrelated, determined by the quantity and quality of available water. Management and allocation of water entails consideration of its unique characteristics as a resource.

Water used for irrigation can be pumped from reserves of groundwater, or abstracted from rivers or bodies of stored surface water. It is applied to crops by flooding, via channels, as a spray or drips from nozzles. Crops also obtain water from precipitation. Water infiltrates into the soil, evaporates, or runs off as surface water. Of the water that infiltrates the soil, some is taken up by plants and some percolates more deeply, recharging groundwater. This water can be polluted with agrochemicals, with salts leached from the soil and with effluent from animal waste.

However, pollution can be attenuated as the water moves through the ground by processes that include sorption, ion exchange, filtration, precipitation and biodegradation. Aquifers can also be sources of pollution. Pollutants can be released into groundwater from pockets of contaminants or natural materials within the aquifer. When river levels are low and groundwater levels are high, groundwater can recharge the levels of surface water, which creates a two-way linkage between resources of surface and groundwater.

It is not easy to control or prevent water use. Many uses of water involve the withdrawal of water from the hydrological system (known as 'extractive' or 'off-stream' use). Typically, only a small proportion of the water withdrawn is consumed. Water consumption is exclusive in its use. Consumed water is retained in plants, animals, or industrial products, so it is not available for other uses. However, most of the water withdrawn is not consumed and it returns to the water system for reuse at a later time and a different location.

Water in return flows can reenter the surface water system further downstream, can percolate into aquifers, or evaporate, returning to the hydrological system in gaseous form. Therefore, water withdrawals are not exclusive within a broad perspective on water use, but only within a narrow location- and time-specific context. Water can also be used in-stream without removal from the hydrological system. Such uses generally entail little or no consumption of water but do affect the location and time at which water is available for consumption by other uses.

Water is a 'bulky' resource. This means that its economic value per unit weight or volume tends to be relatively low. Therefore, its conveyance entails a high cost per unit of volume and is often not economically viable over long distances unless a high marginal value can be obtained. The costs of abstraction, storage and any conveyance tend to be high relative to the low economic value that is placed on the use of an additional unit of water. This can create values for water that are location specific.

A further characteristic of water is that the quantity of supply cannot be readily specified; it is determined by various processes: the flow of water; evaporation from the surface; and percolation into the ground. In the case of surface water, supply is determined largely by the climate. Consequently, the quantity supplied is variable and can be unreliable. This can preclude certain uses of water and affect the value of water in some uses. The quality of water can exclude certain uses, but have no impact on others.

Characteristics of demand for water for irrigation relate to quantity, location, timing and quality. Irrigation generally requires large volumes of water, which can be low in quality. This is in contrast to household use of water, for example, which requires low quantities of water of high quality. The large volumes of water required for irrigation usually have to be transported over some distance to the field. For surface water, canals and pipes can enable conveyance; in the case of groundwater, extraction is provided via tubewells. In terms of timing, demand for irrigation water can extend through the growing season and, where adequate supplies are available, extend into the dry season for multiple cropping.

Peak demand for irrigation water does not usually coincide with peak flows of surface water. This creates the need for storage capacity, which naturally occurring waterbodies or specially constructed dams may provide. Although the quality of water required for irrigation is low, high levels of salinity preclude its use for irrigation, and contaminated supplies can reduce the quality of produce.

Agriculture is implicated in issues that concern water quality. Leaching of effluent from animal wastes, especially from intensive livestock production, can pose a serious water pollution risk. Both return flows of irrigation water and precipitation runoff from arable land can pollute surface water with nutrients, herbicides, pesticides, salts leached from the soil, and sediment.

Agricultural Water Use

Irrigation is a vital component of agricultural production in many developing countries. In 1997-99, irrigated land provided two-fifths of crop production in developing countries, and accounted for about one-fifth of the cultivated area. The divergence in these statistics reflects the high crop yields and multiple cropping that are achieved through irrigation. Developing countries are particularly dependent on irrigation: in 1997-99, 59 percent of cereal production in developing countries was irrigated. Food production in developing countries is increasing in response to the demands of an expanding population and rising prosperity. Some of this demand will be met by increased productivity of rainfed agriculture, some by increased imports, but irrigated agriculture will be a major contributor.

However, the agriculture sector is often criticised for high wastage and inefficient use of water at the point of consumption (i.e. at farm level) encouraged by subsidised low charges for water use or low energy tariffs for pumping. It is often claimed that the charges made for irrigation water, fail to signal the scarcity of the resource to farmers. This situation may persist because of entrenched interests, political problems associated with price reform, practical difficulties in measuring and monitoring water use, and social norms, e.g. perception of water as a free good and access to water as a basic right.

These low charges can have an adverse impact on the effectiveness of irrigation systems and water use. They result in poor maintenance and consequent inefficient operation of existing irrigation systems, limited capacity for improvements or investment in new infrastructure, and waste of water at the farm level. Furthermore it is claimed that the subsidies provided for irrigation water tend to favour the wealthy and thereby exacerbate inequalities in resource access and wealth distribution in rural areas.

Water used for irrigation comes from surface water or groundwater. The use of groundwater for irrigation enables the extension of irrigated area beyond that which surface water alone can support. In addition, it assists with drainage of the soil. Groundwater can supplement surface water during periods of low flow, making surface water available for alternative uses. It is also used as a sole source of irrigation water. For example, in India, more than half of irrigated land is supplied with groundwater, providing one-third of the country's food production.

Groundwater has various advantages over surface water: it can be stored in aquifers for years with little or no evaporative loss; the percolation of aquifer recharge water through the ground attenuates pollution levels (making groundwater particularly suitable as a source of drinking-water, especially in areas with no water treatment facilities); groundwater can be withdrawn near the point of use; and it is available immediately on demand, which enables more timely applications of irrigation water. However, groundwater contains dissolved salts that can be toxic to plants and result in soil salinisation. Groundwater can be combined with surface water to dilute salt concentrations to levels suitable for use in irrigation.

Surface water for irrigation is stored either in natural storage capacity (lakes and wetlands) or artificial capacity created through the construction of dams. Dams are usually constructed for the purposes of water storage for irrigation, hydroelectric power generation, flood control, or any combination of these. However, in the case of dual-purpose dams designed to store water for irrigation and hydroelectric power generation, conflicts can arise because increases in demand for irrigation water in the dry season exceed demand for power. This creates difficulties in the specification of the required storage capacity and the timing of water releases. The situation is yet more complex for dams also designed to provide flood protection. Effective provision of flood control requires storage capacity that is empty, but effective storage of water for hydroelectric power generation and irrigation requires storage capacity that is kept as full as possible.

Despite potential for conflict, provision of storage capacity for irrigation combined with other uses can have advantages. The combined value of storage capacity for multiple purposes may be required in order to make large dam developments economically viable. Moreover, the provision of storage capacity for non-agricultural uses can provide contingency against failure of irrigation schemes to meet predicted uptake and economic returns, e.g. through potential to develop further power generating capacity.

The design and implementation of irrigation projects has traditionally been the domain of engineers and agronomists. In response to a commitment to a more developed approach to water management, a broader multidisciplinary perspective on irrigation is evolving. This approach incorporates social, cultural, environmental and wider economic impacts of irrigation projects. Nevertheless, implementation of this perspective on the ground in the development and management of irrigation projects and programmes remains a persistent challenge. However, this challenge can begin to be addressed by the appropriate deployment of the functional approach to water management advocated here.

Supply of bulk water for irrigation is under pressure from the demands of other water-using sectors, constraints on further water resource development and is compounded by poor maintenance of existing irrigation infrastructure.

Demand for water for non-agricultural uses is increasing in response to economic growth, rising populations and increased urbanisation. Rising urban demands for water (for household and industrial use) pose a particular threat to agriculture because urban demands take priority over rural demands in situations of potential conflict. This is because existing urban supplies are usually polluted, they can be associated with high health risks, new urban supplies have to come from increasingly distant sources, and the economic benefits of urban water supplies exceed those of rural supplies.

Worldwide, withdrawals of water for household and industrial use quadrupled between 1950 and 1995, while withdrawals for irrigation only doubled in the same period. In terms of future demand in developing countries, non-agricultural demand for water is forecast to increase by 100 percent between 1995 and 2025 and agricultural demand to rise by only 12 percent.

Increases in non-agricultural demands for water are coinciding with constraints on further development of new water sources. In combination, these two factors are creating increased water scarcity and they will result inevitably in the transfer of water from agricultural use to higher value household and industrial uses. Urban areas can and do appropriate water supplies from rural areas, resulting in depletion and pollution of surface water resources used by farmers and rural households.

In areas of India and the Philippines, water supplies have been diverted from large irrigated areas, seasonally or permanently, to meet urban demand, without any payment of compensation to farmers for resultant losses in crop production. Increases in household and industrial demand for water are expected to result in increases in the scarcity of water for irrigation

Objectives of Irrigation and Agricultural Development

Governments and donors have traditionally justified allocation of water to agriculture on grounds of food security and rural development. These are examined below, followed by a brief overview of relevant aspects of the international consensus that has emerged in water management policy.

Food security

Irrigation enables greater agricultural production than is achieved with rainfed agriculture. The additional food production obtained with irrigation is essential for food security on a global level, and on a national level for some countries. National food security is attained either through the pursuit of self-sufficiency in food or through a combination of domestic production and imports. Food self-sufficiency was once a widespread objective and some nations still aspire to it. It creates savings in foreign exchange, protects domestic producers and consumers from the fluctuations of world markets, ensures rural food supplies and contributes to a political sense of national security. However, it has disadvantages.

In arid countries, a self-sufficiency policy can increase allocations of water to agriculture at the expense of industrial and household water use, and can contribute to the overextraction of groundwater resources. Moreover, food supplies are vulnerable to extreme weather events, and shortfalls in supply then have to be met through imports, which eat into limited resources of foreign exchange. In response to various factors, which include increased water scarcity, reduced availability of agricultural land, and industrial growth, many countries have moved towards an objective of food security partly enabled by imports. However, successful pursuit of such an objective is reliant on adequate regulation of world trade in foodstuffs, to provide assured imports under fair terms of trade.

Global demand for food is increasing as the population continues to grow and increase in prosperity. Demand pressure is concentrated in developing countries, where demand for agricultural products is forecast to increase at an average rate of 2 percent per year from 1999 to 2030. Food demand is also affected by a shift in diets, which is occurring in developing countries as a result of increased prosperity, urbanisation and changing preferences. Populations in developing countries are tending to consume more livestock products, more fruit and vegetables, and fewer cereals than in the past. Meat consumption in developing countries is projected to increase by 44 percent per capita from 1997/99 to 2030.

Combined with a general shift in animal production from extensive (i.e. grazing) to intensive (i.e. cereal-fed) systems and the low efficiency of meat production, this is creating increased demand for cereals (such as maize) for animal feed. Cereals for animal feed account for half of the projected 70-percent increase in demand for cereals forecast to occur in developing countries between 1997/99 and 2030. Irrigation is used particularly important to produce cereals. For example, almost 60 percent of the cereal production in developing countries in 1997/99 comes from irrigated land. However, it also contributes to meeting increased demand for other foods.

Increased demand for food in developing countries will be offset at a national level, to various extents, by increased agricultural production. A 61-percent increase in annual cereal production is expected to occur in the period 1997/99-2030. With the exceptions of sub-Saharan Africa and Latin America, irrigated agriculture will provide much of this increase. In developing countries collectively, irrigated agriculture will provide 57 percent of the additional 256 million tonnes of cereals that will be produced in 2025 relative to 1995. Irrigation increases agricultural production through both the expansion of cultivable area beyond that possible under rainfed agriculture and higher crop yields.

About 70 percent of the increase in agricultural production that is forecast to occur in developing countries from 2000 to 2030 will come through increased yields, 20 percent through expansion of crop area and 10 percent through increased cropping intensity. Irrigation increases yields not only through reduction or prevention of crop water stress, but also through complementary benefits of combined use of irrigation with high yielding varieties, fertilizers and pesticides. Yields for cereals produced with irrigation exceeded rainfed yields by 115 percent in developing countries collectively and by 150 percent in sub-Saharan Africa and West Asia/North Africa in 1995. Although yields for irrigated cereal production in developing countries are increasing by 1.2 percent per year, it is at a reduced rate relative to 1982-1995.

Increases in yields for irrigated cereals in developing countries are expected to be of a similar proportion to increases in yields for rainfed cereals in the period 1997/99-2030. However, higher initial yields for irrigated cereals will result in greater absolute increases over this period. For developing countries collectively, average weighted yields for irrigated cereal production are expected to increase by 1.4 tonnes/ ha, compared with an increase of 0.5 tonnes/ha for rainfed cereals between 1997/99 and 2030. Irrigated agriculture is thereby forecast to contribute significantly to increased future food production through both high and increasing crop yields.

In addition to increasing productivity, irrigation also enables expansion of the area under cultivation. In 1997/99, irrigation was used on 21 percent of arable land in developing countries collectively, though this was subject to considerable regional variation. In South and East Asia, irrigation was used on 39 and 31 percent of arable land, in the Near East and North Africa, 30 percent of arable land was irrigated, and in sub-Saharan Africa and Latin America, irrigation was used on only 2 and 9 percent of the arable area, respectively. Expansion of the area under irrigation is expected to be concentrated in developing countries.

Absolute increases in irrigated arable area for the period 1997/99-2030 are forecast to be greatest in Asia. In sub-Saharan Africa and Latin America, expansion of the irrigated arable area is expected to be low in absolute terms, though these represent large proportionate increases. In the period 1962-1998, the area under irrigation in developing countries increased at an average rate of 2 percent per year, adding a total of 100 million ha to the area under cultivation. However, the net increase in irrigated area in developing countries from 2000 to 2030 is expected to be 60 percent less than the net increase achieved for the period 1960-2000. The forecast growth rate in irrigated area (0.6 percent per year) is one-third of that achieved in the period 1960-2000. In developing countries collectively, the slowdown in the development of irrigated arable land will be countered to some extent by expansion in the arable area under rainfed crop production. Consequently, the share of total cereal area under irrigation in 2030 will remain relatively unchanged from 1997/99. In developing countries collectively, 22 percent of the arable area will be irrigated.

Climate change is expected to affect agricultural production in developing countries, particularly through increases in temperature in arid regions and greater variability in the climate. It will tend to increase local fluctuations in crop production and food supplies, particularly affecting food supplies and the incomes of poor people, and to increase national vulnerability to food insecurity. In certain regions, the effects could be significant even in the next few decades. For example, climate change could cause a 2-3-percent decline in cereal production in Africa by 2020 or 2030. Assuming that other factors remain constant, this would increase the number of people at risk of hunger by 10 million.

Demand for food is not met solely by domestic production in many developing countries; imports of food are required to varying extents. In 1997/99, cereal production represented 91 percent of demand for cereals in developing countries. However, this aggregation hides regional extremes. In the Near East and North Africa, domestic cereal production represented 63 percent of demand. In sub-Saharan Africa and Latin America, it represented 82 and 88 percent of demand, respectively, and in East and South Asia, production met 95 and 102 percent of demand. Reliance on imports is forecast to increase.

Poverty Alleviation

In an appropriate environment and with suitable planning, investment in irrigation schemes can alleviate poverty both directly and indirectly through stimulation of the rural economy. Indeed, the purpose of many large scale schemes associated with the Green Revolution in

Asia was more to do with addressing food security and poverty targets rather than direct commercial returns.. This notion and practice persists. The IFAD "Report on Rural Poverty 2001" is clear in stating that irrigation schemes have direct benefits for poor people, given the required policy and institutional environment. Even if irrigation is not specifically targeted at poor beneficiaries, irrigation stimulates the agriculture sector of the rural economy indirectly through increased demand for agricultural inputs and the marketing of additional produce.

Increased incomes in farming communities can create demand for non-agricultural goods and services, many of which are marketed only locally and can be supplied by resource-poor individuals. The resultant stimulation of non-farm incomes can help to reduce absolute poverty in rural areas in the long term, and it can reduce relative poverty as long as the prevailing asset distribution is not too skewed.

Increased food production from irrigated agriculture can confer nutritional benefits for farmers, their families and the local population (through increased food supplies). Irrigation can enable multiple cropping, which can smooth seasonal shortfalls in food supply and encourage the production of crops that contribute towards a more varied and nutritious diet. Improved nutrition can enhance quality of life, reduce illness, increase labour productivity, and improve the performance of children at school.

Irrigated agriculture can also benefit the urban poor by keeping food prices low despite growing demand from increasing populations. Indeed, continuation of the current decline in irrigation investment could eventually cause an increase in world cereal prices food prices, which would affect the poor in particular as a large proportion of their income goes on food.

However, irrigation can have a negative impact on the health of rural households through exposure to parasitic infections and to diseases transmitted by water-related vectors such as malaria. Moreover, in an inappropriate environment, e.g. where land is not evenly distributed, economic benefits of irrigation may be received predominantly by wealthy farmers and reinforce inequalities in the distribution of resources and wealth. The policy and institutional environments play critical roles in determining whether irrigation has positive impacts for poor people.

Water as an Economic Good

An international consensus in water management has emerged, based on growing concerns about efficiency in the use of government and donor resources, disappointing outcomes from past efforts, and greater awareness of environmental issues. These concerns are manifest in for example, the water policy of the European Community, which promotes the use of water pricing and charging as a means of enhancing the sustainability of water resources, and on integrating economics into planning and decision-making. The policy consensus has also been shaped by a reorientation of the development cooperation agenda that has resulted, among other issues, in greater focus on institutional reform, participation, and involvement of civil society and the private sector.

Three agreements lie at the core of the consensus concerning water policy:

(i) a set of key recommendations (known as the "Dublin Principles") agreed at the International Conference on Water and the Environment (1992);

(ii) Chapter 18 (on freshwater resources) of Agenda 21, the action plan agreed upon at the United Nations Conference on Environment and Development held in Rio de Janeiro in 1992 and incorporating adoption of the Dublin Principles for water resources management in rural contexts; and

(iii) the World Summit on Sustainable Development held in Johannesburg in 2002, which reaffirmed the 1992 "Dublin Principles" and highlighted water availability as a key concern and objective.

The consensus adopts an integrated approach to water resources and a multisectoral view of water use on at least a catchment scale. Water management is considered in relation to issues of economic efficiency, environmental protection, sustainability, and the needs of marginalised and poor people. Decision-making should involve the participation of users, particularly women, and should be driven by the needs of the community. Investments in the water sector are required to be economically efficient, socially acceptable and financially sustainable.

Despite the consensus on water policy, there is considerable debate on the practical implementation of any reforms. For example, while reforms in economic pricing of irrigation water have been proposed, the political economy of pricing policy reforms suggests a rather complex process in evidence. It is recognised that the main management challenge is not a vision of integrated water resources management but a "pragmatic but principled" approach that respects principles of efficiency, equity and sustainability, but recognises that water resources management is intensely political, and that reform requires the articulation of prioritised, sequenced, practical and patient interventions.

Sustainability is a key aspect of water management decisions. However, the term sustainability is open to different interpretations, and these affect how it may be operationalised. The differences arise through use of a flexible or stringent interpretation of sustainability, described as weak and strong sustainability, respectively.

Sustainability requires that the stock of capital that is available for future generations be equivalent to that available at present. Here, the term capital refers to the overall stock of materials and information that generates goods and services that enhance social welfare. Capital can be subdivided according to whether it is depleted by the production of goods and services, as follows:

— Capital caused by human activities (e.g. factories, roads and houses). This can be increased or decreased at discretion.

— Critical natural capital (e.g. ozone layer, biodiversity and water). This is essential to human life and cannot be replaced by or substituted for with human-induced capital.

— Non-critical natural capital, which includes some renewable natural resources and some finite mineral resources. This can be wholly or partly replaced or substituted by human-induced capital.

Weak sustainability requires that the total stock of capital, human-induced and natural, be maintained, and it assumes substitutability between the two types of capital. As a natural resource becomes depleted, the price increases; this encourages more efficient resource use, substitution with other goods, and technological advancement. However, complete substitution is not necessarily practicable or possible, for example, because of the absence of substitutes for some forms of capital, and inadequacy of substitutes.

Strong sustainability requires that the total stock of natural capital not be depleted. Natural and human-induced capital are regarded as complements not substitutes and stocks of both must be maintained. Consequently, activities are required to conserve the natural environment, or to ensure that any losses incurred are replaced or compensated for fully in physical terms through 'shadow projects'.

The use of a criterion of strong sustainability is likely to result in the wholesale rejection of development projects as most of these impinge to some degree on the environment. However, this rejection can be overcome by employing suites of projects that are designed to have elements that generate net environmental benefits.

Adoption of such an approach enables market-oriented decision-making to persist even under stringent sustainability requirements. The wetland mitigation policy of the United States of America provides an example of a strong sustainability requirement. The policy requires that any loss of wetland be compensated for by an alternative wetland of equal physical quality. However, a number of problems have been encountered in implementing the policy. These include the definition of a suitable measure of the physical quality of wetlands, and issues that relate to the locality and interactions with the landscape.

The emerging approach to water governance is seeking to adopt a stronger sustainability approach, one guided by principles of stewardship, equity and accountability. The result will be to constrain the mindset and market mechanisms that treat water as a commodity in its various functions and seek to establish an efficient allocation of water among competing end uses. Efficiency is a necessary but not sufficient condition for sustainability, but exactly how constraining sustainability standards ought to be remains an open scientific and policy question.

Economics of Water Allocation

Water resources provide important commodity and environmental benefits to society. Any particular use of water will be associated with opportunity costs, which are the benefits foregone from possible alternative uses of the resource. Decision-makers are faced with balancing, for example, water demands from agricultural irrigation for food production with the desire to preserve wetlands for fish and wildlife habitat. Economics contributes towards improved allocations by informing decision-makers of the full social costs of water use and the full social benefits of the goods and services that water provides. The main approaches that form the methodological basis for strategic economic appraisal are cost-benefit analysis and cost-effectiveness analysis.

Cost-benefit Analysis

Cost-benefit analysis is carried out in order to compare the economic efficiency implication of alternative actions. The benefits from an action are contrasted with the associated costs within a common analytical framework. The benefits and costs are usually measured physically in widely differing units; comparison is enabled through use of the common numeraire of money. The benefits and costs of each option are determined relative to the common scenario that would prevail if no action were taken. The net benefit of each option is given by the difference between the costs and benefits. The most economically efficient option is that with the highest present value of net benefit, i.e. net present value (NPV); economic efficiency requires selection of the option with maximum NPV. Options are economically viable only where the NPV that they generate is positive.

Cost-benefit analysis provides a rational and systematic framework for assessing alternative management and policy options. It entails identification and economic valuation of all positive and negative effects of alternative options. This involves the translation of all benefits and costs into monetary terms, including where possible, non-marketed environmental, social and other impacts. It is based on the underlying assumption that individual preferences should determine the allocation of resources among competing uses in society.

Selection of cost-benefit analysis as a decision-support tool is at the discretion of analysts and policy- and decision-makers. It is the responsibility of analysts to ensure that the underlying assumptions of a cost-benefit analysis are appropriate to a specific situation and that the results are valid and reliable. Cost-benefit analysis reflects a specific paradigm that is considered more or less appropriate in different domains of decision-making. The appropriateness depends on culturally determined beliefs, norms and values, regarding, for example, the legitimacy of the social-political organisation of decision-making, public consultation and the process of value elicitation.

Cost-effectiveness Analysis

Cost-effectiveness analysis (least cost analysis) is used to identify the most cost-effective option for achieving a pre-set objective or criterion. The relevant objective is set, options for achieving it are identified, and the most cost-effective option is identified as that with the lowest present value of costs. It is assumed implicitly that the benefits of meeting the goal outweigh the cost and that the action is therefore economically viable.

Cost-effectiveness analysis is suitable for use in situations where valid and reliable estimation of the benefits of alternative options is not feasible. This is particularly relevant to actions that involve environmental change. Instead of attempting to identify and value the benefits, the most cost-effective means of achieving a desired objective is identified.

For example, cost-effectiveness analysis is suited to situations where clear and defensible conservation targets or other environmental goals exist that can be measured in terms of biophysical units such as minimum water quality standards. It can also be used to identify the most effective option for a fixed amount of funding that has been allocated to achieve a policy objective. The drawback of cost-effectiveness analysis is that it does not identify the benefits of actions or the willingness of society to pay for improvements in environmental quality, which are important considerations in many decision contexts. For these reasons, cost-benefit analysis is, where practicable, the preferred tool for decision support.

Economically Efficient Allocation

The focus on economic efficiency as the primary objective in the development and allocation of water resources is because of its importance as a social objective; efficiency values having viable meaning in resolving conflicts and assessing the opportunity costs of pursuing alternative uses. Although economically efficient allocation of irrigation water is rarely attained in practice, analysis of economic efficiency provides a useful point of reference for understanding causes of inefficient allocation and mechanisms for improving the overall economic performance of irrigated production. At the outset, economic efficiency needs to be distinguished from the various technical definitions of efficiencies associated with irrigation. However the two types of efficiency are related in the sense that both seak to maximise the productivity of water in terms of output per cubic metre of water.

Economically efficient allocation of water is desirable to the extent that it maximises the welfare that society obtains from available water resources. Welfare in this context refers to the economic well-being of society and is determined by the aggregate well-being of its individual citizens. Economically efficient allocation maximises the value of water across all sectors of the economy. This is achieved through the allocation of water to uses that are of high value to society and away from uses with low value.

Efficient allocation occurs in a competitive, freely functioning market when supply is in equilibrium with demand. Under these conditions, the marginal cost of the supply of water is equal to the marginal benefit of the use of water. The marginal benefit and marginal cost are the same across all uses and equate with the market price. However, where there are distortionary constraints, such as subsidies or taxes, the maximisation procedure will result in a second-best efficient allocation.

A feature of economically efficient allocation is that no reallocation can make anyone better off without making at least on person worse off, a condition that is described as "Pareto optimal". The relative efficiency of alternative allocations can be analysed with respect to this, i.e. in terms of whether they provide a "Pareto improvement". A change in allocation is considered desirable if at least one person gains in welfare and no one loses. However, this criterion proves too stringent in practice as few changes can be made in the real world that do not reduce the well-being of others. For this reason, an adaptation is usually employed; this is described as a 'potential Pareto improvement' or the Kaldor-Hicks criterion.

A change in allocation is considered desirable if those individuals who gain from the change can hypothetically compensate those who lose and still be better off than they were previously. It is anticipated that compensation does not take place, owing to difficulties of identifying and compensating all necessary individuals. The criterion of potential Pareto improvement forms the basis of cost-benefit analysis, which is used to analyse the relative economic efficiency of alternative courses of action.

Although economic efficiency is an important factor, there are additional economic issues that decision-makers need to consider. Two of these issues are the distribution of costs and benefits across society and their distribution across generations. In terms of the former, neither the equity implications of an allocation nor the equity of the prevailing distribution of wealth are considered in analysis of economic efficiency.

Focusing first on the equity implications of an allocation, costs and benefits are usually specified using values that are representative of the whole of society. However, the costs and benefits may not be borne equally by society; they may be concentrated in specific geographical areas. These differences may also correlate with differences in income borne by sections of society: environmental costs are often borne disproportionately by low-income sections of society. Such disparities can be incorporated into analysis through studies of costs and benefits for separate sections of society, though this adds to the information requirements and the demands of the analysis.

The prevailing distribution of wealth is usually assumed to be a given in analysis of economic efficiency. Equal weight is given implicitly to costs and benefits experienced by all members of society. However, circumstances arise where it is socially desirable to alter the distribution of wealth in the pursuit of greater equity. This can be incorporated into the analysis through the use of distributional weights.

Weights are assigned to costs and benefits according to the section of society that they accrue to and the desired redistribution of wealth. For example, high weights can be applied to benefits that accrue to poor sections of society and low weights to benefits for the rich. Application of this procedure is challenging because of the difficulties of identifying the distribution of costs and benefits within society and of specifying appropriate weights, which is subjective. In the past, it has usually been considered more appropriate for decision-makers to consider prevailing inequalities separately from analysis of economic efficiency.

In terms of equity in the distribution of costs and benefits over time, it is argued that economic analysis commonly favours consumption in the present at the expense of future generations. Analysis of economic efficiency addresses the distribution of costs and benefits over time through the use of discounting: all costs and benefits are converted into present values using a rate of time preference. The discount rate used is intended to represent society's preference for consumption in the present over the future. The rate is commonly prescribed by government agencies and is typically positive and less than private discount rates. On the one hand, it can be argued that government has a responsibility for the interests of both current and future generations and that it sets the social discount rate accordingly. One the other hand, it is argued that the rates used are too high, placing low weight on consumption in the future and discriminating against future generations.

A further argument concerns the use of a zero discount rate, which would place equal weight on the interests of all generations. However, its application would result in rejection of all use of non-renewable resources or any irreversible developments. It could also result in rejection of investments that might be of great value to future generations, through creation of wealth or new technologies. Concerns about future generations can be addressed better if they are considered separately from analysis of economic efficiency, through, for example, application of safe minimum standards.

Inefficient Allocation

Although water resources perform many functions and have important socio-economic values, water is in many respects a classic non-marketed resource. Even in its use as a tradeable commodity, market prices are not generally available. The reasons why water has no price are often related to the historical, socio-cultural and institutional context in which water is used and managed. In addition, although water can be captured and shared, water flows can also be recycled. This often makes it difficult to break water down into marketable proportions.

An important cause of this economically inefficient water use (where costs outweigh benefits) is the failure of institutions involvement with the allocation and management of water. 'Failure' refers here to institutions where 'they induce or favour decisions that lead society away or prevent society from achieving socially optimal resource allocations'. Sources of institutional failure include markets, policies, and political and administrative factors. They derive from a fundamental failure of information or lack of understanding of the multitude of values that may be associated with water resources.

Market Failure

Although markets can achieve economically efficient allocation, they are commonly unable to do so. Described as market failure, this occurs through an 'inability of the market to lead the economic process towards the social optimum'. Market failure can occur through the non-existence of markets, their failure to communicate necessary information, restricted operation of markets, and inadequate institutions or regulations.

Activities can impose losses or gains in welfare on individuals other than those engaged in the activities. If these losses or gains go uncompensated or unpaid for, they are described as externalities. Externalities are not incorporated into market prices, so are not accounted for in market-based allocation. This results in socially suboptimal resource allocation and market failure. Return flows of water are an example where both positive and negative externalities can be generated. Negative externalities arise where pollution from water use imposes additional treatment costs on downstream users. In the context of irrigation, drainage water from fields often carries high levels of agrochemical pollutants, which can lead to losses of aquatic habitats downstream.

Optimal allocation of water requires that supply costs be increased to reflect the costs of mitigating the negative externalities. Positive externalities are also generated by return flows, which form a vital element of many hydrologic systems. Irrigation often performs a secondary function in that it recharges aquifer systems. Such external effects can mean that while farm-level water use efficiency may be apparently low, at the catchment level, water use efficiency of irrigation may be much greater. Thus, improvements in efficiency at the farm level may be at a cost of overall efficiency of the hydrological system.

By definition, a public good can be enjoyed without diminishing the supply and others cannot be excluded from its use. As a result of the non-rivalry characteristics, demand for public goods is collective: it is the sum of the separate demands of individuals for the good. Although some uses of water tend towards being rival in consumption, e.g. agricultural, residential or industrial uses, others such as recreational and aesthetic uses are non-rival. Thus, water supply has often been exposed to 'open access' pressures, with a lack of enforceable property rights allowing unrestricted depletion of the resource. Furthermore, even where water resources are privately owned, many of the benefits they provide may be off-site, and may not accrue to the owner. The lack of a market for these benefits limits the incentive to maintain the resource, as the private benefits derived by the owner do not reflect the full benefits to society. Most commonly, the non-traded nature of public goods hides them from market-based decision-making, which results in market failure.

The preference of individuals for consumption in the present rather than the future is understood to usually exceed that of society. Where allocations are determined based on the discount rate of an individual decision-maker, this is likely to give less weight to long-term costs and benefits. Typically, this translates into the selection of courses of action that are of short-term net benefit and rejection of those that are of net benefit only in the longer term. A particular concern is the favourable consideration that is given to courses of action that yield net benefits in the short term, but incur substantial costs on society in the long term.

The supply of irrigation water is often controlled by only one agency, a situation described as a monopoly. Under these conditions, the supply of water is not subject to market competition. The supplier determines the price and quantity of water supplies. This can result in inefficient allocations and is a source of market failure. For example, a monopolistic supplier may elect to allocate water between farmers in a manner that does not make the maximum contribution to social welfare. Similarly, the supplier may set the water supply at a level that exceeds the optimum for society in order to maximise profits.

Property rights are the characteristics that define the rights and duties associated with use of a particular source of water. The nature of these property rights can determine the efficiency with which a source of water is used. Particular types of property rights regime result in market failure. Property rights regimes can be considered in terms of four types: private, common, state and open-access.

With private property rights, a private individual owns the resource and has the right to use, benefit from, and sell the resource, subject to a duty to refrain from socially unacceptable activities. These rights are subject to state regulation and protection, which is required for private property to exist and to exclude unentitled individuals from its use. However, the rights can be eroded in circumstances where resources supply public and private goods jointly. In order to maximise profits, the owner of a private resource manages the resource like an asset: its use is allocated over time such that the total present value is maximised. The owner delays use of the resource because of the expectation of being the only beneficiary of the resource in the future.

With common property rights, a group of individuals owns and manages the resource. The individuals in this group have specified rights and duties regarding the resource, and enforced rules exclude other individuals from its use. Common property ownership of resources tends to play a larger role in developing countries than in the industrialised countries. Communities that have the characteristics required for successful common management of resources are more prevalent in developing countries: communities that are small, relatively immobile, close knit, and that have a common ethos and shared beliefs. Moreover, common property resources can have an important role in these communities as they can aid the distribution of wealth and provide a means for reducing the marginalisation of poor, e.g. provision for the harvesting of fish and plants from communally owned wetlands by low-income households.

With state property, the state owns the resource. The state may allow individuals to use the resource, but according to its rules and under its regulation. In an open-access situation, no property rights are assigned to the resource, which results in open access to the resource for all potential users. Water from both underground and surface sources is often an open access resource. Use of the resource is subject to neither exclusion nor regulation. Individuals have complete autonomy in its use.

The property rights of resources are often held in combinations of the above regimes and can alter with a change in situation. The efficiency of resource use under these regimes is based on four determining conditions:

— full specification of ownership and entitlement to the resource (universality);

— accrual of all benefits and costs exclusively to the entitled individual (exclusivity);

— exchange of property rights in voluntary transactions (transferability);

— penalties that prevent individuals from encroaching or taking property rights without prior agreement (enforceability).

Policy and Institutional Failure

Allocation of water can be socially suboptimal not only through the failure of markets, but also through failure of government policy and associated instiututional arrangements. Policy failure occurs where government regulatory instruments or government policies create market price distortions that make it economically rational for individuals to use resources in a socially suboptimal manner. Sectoral policies for agriculture and the environment, and for other sectors such as employment and taxation, can encourage suboptimal resource use and allocation. Moreover, government interventions intended to correct for market failure can result unwittingly in greater degradation and depletion of environmental resources if the regulatory environment is not suitably 'joined up'.

Failure in sectoral policies can arise through inadequate consideration of impacts on other sectors, particularly with regard to the environment. Bias in the formulation of policy towards sectors that exercise strong economic and political power can further marginalise the concerns of other sectors. Agriculture provides various examples of policy failure, such as subsidised irrigation and land improvement. A policy of subsidised supplies of irrigation water provides farmers with water at a price that is less than the marginal cost. Farmers respond to the low price by using quantities of water that exceed the social optimum. In an extreme case, this can result in overabstraction of aquifers and waterlogging of the land. An alternative example of agricultural policy failure is the provision of subsidies to drain land and to divert surface water, both of which can contribute to the degradation or destruction of wetland sites.

Political, administrative and other institutional factors can distort market signals in a manner that encourages socially suboptimal resource use. Pressure to gain re-election can lead governments to over-supply state services for example. At the extreme, this can favour excessive economic development at the expense of resource conservation. Political failure can also occur through lack of government intervention, e.g. in response to market failure. Administrative failure refers to a range of problems within the organisation of government at the various levels, leading to inadequate policy implementation. Examples include: rigidities due to entrenched traditional divisions of labour within administrative organisations, and insufficient integration between agencies and departments. Other institutional failures include inadequate availability of information for policy-makers, and poor communication between rural electorates and urban-based central government. These policy and institutional failures are generic and could apply to most government interventions. The significance for irrigaton is the scope and depth of government involvement in irrigation.

Water Allocation Policy

Economic efficiency and equity are important considerations in the allocation of water. Greater efficiency is required in the face of increasing water scarcity, and equity is a concern because of the importance of water to the livelihoods and well-being of rural communities in particular. It is possible to derive a broad classification of policy measures that are relevant to managing resources within the boundaries of a nation. The measures include the redefinition of property rights and investment policies. One thing to be kept in mind is that many of the large public irrigation schemes that were promoted as part of the green revolution, particularly in Asia, were designed to target poor rural communities and as such were never oriented to maximise economic output, rather to guarantee production of food staples.

The proper pricing of inputs (such as raw water) and outputs (such as agricultural irrigation products) can be viewed as a form of property right designation, while command and control measures are also means of defining property rights or modifying existing ones. Once a regime of property rights has been established, the proper pricing of a resource requires that it be priced at least at marginal private costs, and preferably at marginal social cost (especially in the longer term and where output prices are below private production costs).

As pricing of water affects the allocation decisions of those with competing wants, then by correctly pricing water, efficient allocation of water is achieved. However, the standard economic efficiency (marginal) cost pricing result is sometimes problematic as regards the specification of production technology. In the water supply sector, inputs to production are often not perfectly divisible. Investments often require large lumps of capital (e.g. for dams and reservoirs). In such cases, marginal cost pricing to achieve economic efficiency requires some form of intervention.

Water allocation systems

Water allocation systems differ in the extent to which they address efficiency and equity iobjectives. The various systems can be compared according to several criteria. These criteria include:

- *Flexibility in allocation of supplies*: allocation requires flexibility such that supplies can be shifted between uses and sectors, as demand changes, so as to achieve efficiency.
- *Security of tenure for users:* established users require security of tenure if they are to be expected to take the necessary measures to use the resource efficiently. Although this may conflict with flexibility, problems should not arise if sufficient water reserves are available to meet unexpected demands.
- *Payment of real opportunity costs of water by users:* users should pay the real opportunity costs of their use, so that other demand or external effects are internalised.
- *Predictability of the allocation outcome*: in order to achieve the best allocation and minimise uncertainty minimised, the outcome of the allocation process needs to be predictable.
- *Equity in the allocation process*: users should perceive the allocation process to be equitable.
- *Political and public acceptability:* the allocation should serve the various political and public values and objectives, thereby making it acceptable to the groups in society.
- *Efficacy in achieving desired policy goals*: the form of allocation should change an existing undesirable

situation towards one where the desired policy goals are achieved.

— *Administrative feasibility and sustainability:* the allocation mechanism must be practicable, adaptable and allow an increasing effect of policy.

Water allocation systems range from government-controlled to market-led systems, and combinations of the two. The prevailing institutional frameworks and the water resources infrastructure influence the precise nature of allocation systems. However, they commonly fall into one of only a small number of categories: public allocation, market-based allocation and user-based allocation.

Public allocation of water is determined by the state. It is used for intersectoral allocation of water, as the state usually is the only institution that has jurisdiction over all sectors of the economy and because allocation of water is considered too important to leave to the mercy the market. The state can control allocation within sectors through, for example, granting permits for water abstraction. In agriculture, the state commonly administers allocation of water to large-scale irrigation schemes and to sections within the schemes. Distributions can be based on historical allocations or political influence. The state is less commonly involved in allocation at the farm and field levels. Under such allocations, the price of water is usually subsidised, low and charged on a fl at-rate or fixed-charge basis.

Public allocation of water enables pursuit of objectives other than economic efficiency, such as equity and environmental protection. Some aspects of water allocation lend it particularly to state control. The economies of scale and the high levels of investment required for infrastructure readily create monopolistic supply and consequently a need for regulation. The joint provision of goods and services by water resources and the provision of public goods also suit state-controlled allocation. Many water resources are open access, so require state regulation of use. The interdependence of surface water and groundwater resources can require regulation of abstractions to prevent depletion of surface water and groundwater supplies. Finally, the essential role of water in meeting basic needs can require state control of allocation under conditions of drought.

However, there are problems associated with public allocation of water. These arise through poor management of infrastructure and inadequate development, inadequate implementation of regulations, and subsidised prices. Public allocation usually precludes user participation, a key objective in the international consensus for water resources policy. Moreover, the institutions engaged in public allocation are typically sector-oriented, which fosters neither the integrated nor the flexible management required for effective intersectoral allocation. Thus, where public allocation is necessary, efforts are required to limit the potential deficiencies.

User-based allocation of water is undertaken through collective management of water sources, supplying water for either collective or individual use. Examples include farmer-managed irrigation systems and village-managed local water supplies. User-based allocation requires established rights to water use and an appropriate institutional framework that has the capacity and strength to determine and regulate use. Its effectiveness is determined by the characteristics of the community and its institutions, and by the extent to which social norms influence water use.

User-based allocations have the advantages that they are informed by knowledge and understanding of the needs of the local population, and can be flexible and responsive. Effectively operated systems can allocate water efficiently and address equity concerns. User-based allocations are sustainable, politically acceptable and are supported by the consensus in international policy. However, effective allocation is dependent on the existence of a strong and transparent institutional framework, which is not always present.

Market-based allocation is determined by transactions in water use, at prices determined by the forces of supply and demand. It encourages greater efficiency in the use of water, and flexible and responsive allocation. Such allocation is reliant on effective operation of the market. Markets for water are often hampered from achieving this by subsidies, few suppliers and inadequate information. Moreover, they are distorted by externalities: uncompensated costs and benefits of water use that are imposed on others. These costs and benefits are not internalised into the price of water, which results in pervasive socially undesirable outcomes.

Where it is judged desirable to establish a market for water, trade in the rights to water use is generally more acceptable than volumetric pricing as farmers view the latter to entail appropriation of their prior rights in water use. Such establishment of a market requires clearly defined rights for water use, specification of the initial allocation of rights, the necessary institutional framework and the infrastructure required to enable trade in water. In this sense the basis used for the pricing is an important determinant of the acceptability.

Pricing and cost recovery in the irrigation sector

Irrigation water is commonly priced volumetrically or using a fl at rate or fixed charge. A special case of volumetric pricing is marginal cost pricing which requires metering of water use and the necessary administrative capacity. A review of World Bank policy and practices recommends that for efficient use, irrigation water is priced volumetrically, based on opportunity costs. Where this is not feasible, water can be charged for at a fl at rate or a fixed charge.

The charges are based not on the amount of water used but on other variables such as the land area, value of landholding, crop output, or non-irrigation inputs. The most common form of charging is based on land area, as this is easy to administer and suited to continuous flow irrigation. However, the actual preview of charging for irrigation water is not necessarily consistent with economic expectations and straightforward notions of price and costs.

Nevertheless, farmers place high marginal values on irrigation water, often a number of times higher than the charges actually imposed and increases in charges may not affect demand where these marginal values are so high. On the other hand, underpricing can be expected to encourage wastage of water, poor maintenance of irrigation systems and inefficient applications of water, resulting in reduced agricultural output.

With regard to private irrigation schemes, while irrigators have to meet the full financial costs of private irrigation schemes, it can be agreed that they rarely face the opportunity costs of water use. However, public irrigation schemes throughout the world have been subsidised to such an extent that charges rarely cover even operation and maintenance costs. Recovery of at least these costs is needed to enable the maintenance of irrigation systems, which is crucial to improved irrigation performance. Nevertheless, cost recovery can achieve such improvements only if the associated revenues are applied to improvements in the system.

In public schemes, farmers sometimes do face a restricted measure of opportunity costs, which can arise implicitly as a consequence of water rationing. However, this is likely to underestimate the true opportunity costs significantly. Simulation analysis has indicated large differences in the economic losses arising from water shortages between market systems and public allocation systems. Market systems were also found to be greatly superior in terms of the equity of distribution of the losses resulting from water shortages.

There are further lessons that are important to the successful reform of water resource allocation policy. Although conventional economic wisdom suggests that users should pay the full economic costs of water supply, pursuit of such an aim is impractical and unhelpful in most developing-country contexts. Users commonly resent paying prices for water that exceed financial costs of supply, and object to paying for water supplies that were previously free. In order to be acceptable, tariffs need to be set on a basis that is understandable, transparent, legitimate, and that stimulates accountability. The challenge in irrigated agriculture is to ensure that farmers take into account the opportunity costs of water use and that institutional arrangements are in place to ensure that water moves to higher value uses.

The incorporation of opportunity costs into water tariffs is not a straightforward task for a number of reasons: the information requirements are considerable; such charges would be objected to on the grounds that they entail appropriation of current users' 'property rights'; and farmers would have to pay substantially more than the cost of service provision, which may be politically unacceptable.

Emerging international experience suggests that the appropriate approach to ensure that users consider the scarcity value of water is to clarify property rights and facilitate the leasing and trading of these rights. "Getting prices right" is important. Nevertheless, the lessons from experiences with pricing irrigation water suggest that water allocation methods are sensitive to physical, social, institutional and political settings, thus making it necessary to design allocation mechanisms accordingly.

Pricing, Opportunity Costs and Economic Benefits

As water becomes increasingly scarce, the legitimacy of treating it as a 'free' resource arises. Arguably, the absence of pricing, as well as the lack of cost effective recovery, has been a major determinant of inefficient and excessive agricultural use of water. In response to these problems, many countries and water management agencies are turning to water pricing mechanisms to allocate water. However, economists disagree on the appropriate means and methods for pricing water and on the notion of an optimal water pricing policy.

The methods used to price water and the performance of these is dependent on the physical, social, institutional and political context. Nevertheless, as the pricing of water affects the allocation decisions of those with competing wants, then correct pricing can led to efficient allocation. In this context, pricing refers to the introduction of amended financial charges in situations where water was previously free or underpriced, or to the consideration of the economic value of water in decision-making through use of an appraisal and accounting procedure, such as cost-benefit analysis.

Because water resources are often non-marketed, it is extremely important to ensure that, where possible, the 'true' economic value of these resources is accounted for when making investments, or decisions concerning water and environmental policy. 'Accounting' or 'shadow prices', determined through the economic valuation of water resources, are employed in such decision-making in place of market prices. Unless water resources are priced correctly, and those prices internalised in decisions, distortions arise. These bias investment and policy decisions against concerns about water resource depletion and degradation, resulting in misallocation of resources and suboptimal social welfare.

The opportunity costs of water resource use and the economic value of the benefits can be compared in terms of whether the use is economically sustainable or socially optimal. The characteristics of the agriculture sector and its relationship to the water system require careful consideration when designing instruments to internalise the opportunity costs in water use and/or to change water use behaviour. These include:

- the nature and complexity of water use and/or pollution;
- geographical/location characteristics;
- characteristics of the target group;
- market characteristics.

However, given the criteria discussed earlier for comparing allocation mechanisms, no general guidelines exist as to how to price water. The right design depends upon the specific characteristics of the agriculture sector and the aim one has in mind of pricing water in the first place.

Marginal opportunity cost (MOC) is an important and useful tool for conceptualising and measuring the physical effects of resource depletion and degradation in economic terms. MOC seeks to measure the full societal cost of an action or policy option that employs a natural resource such as water. Economically efficient resource management requires that the price that users pay for resource use should equate with the MOC. Where the price is less than the MOC, then the resource is overconsumed or overutilised. A price that is higher than the MOC results in the resource being underconsumed or underutilised. Sustainable management is achievable through sustainability pricing, which also includes a premium to cover the costs that accrue from any resource depletion.

The concept of opportunity cost is used to refer to the value of a resource in its best alternative use, i.e. other than the purpose being considered. This is the cost to society of use of the resource. It is considered in terms of a change at the margin, i.e. the MOC, because management decisions usually entail relatively small changes in resource use. MOC comprises three components.

The first component is the direct economic costs of water abstraction, such as the costs of labour, equipment and materials used for abstraction. Such costs require adjustment for any subsidies, taxation and market imperfections in order to reflect true opportunity costs. These costs vary with the difficulty of extraction.

The second part of MOC is external costs that arise from water use. This is the net value of any losses and gains in welfare that water use imposes on individuals other than those engaged in the activity. External costs arise because changes in one component of the natural resource base affect other components and the efficiency with which other activities can be conducted. Costs that occur in the future require discounting in order to make them commensurate with present day costs. Although information on marginal external costs is difficult to obtain and often imprecise or incomplete, useful approximations are possible. It is the external costs that arise from unsustainable resource use that are of particular interest.

The final component of MOC is relevant for non-renewable resources. Where such resources experience a positive rate of exploitation, then use of a unit of the resource results in its non-availability for future use. A scarcity premium can be placed on the resource, its magnitude depending on: the size of the resource stock relative to the rate of exploitation; the strength of future demand relative to present demand; the availability and cost of future substitutes; and the discount rate. This scarcity premium is known as the user cost and relates to the value of the opportunity foregone by exploiting and using the resource in the present period rather than at sometime in the future. It also incorporates increases in the costs of future resource use and exploitation that occur as a consequence of current use and exploitation. Marginal user costs also apply to non-sustainable use of renewable resources.

The user cost of non-renewable water resources is often ignored, especially where water resources are treated as an open-access resource and users behave in an individually competitive manner. This can happen in situations where property rights are ill defined or not enforced. Use of the water is then governed by the law of capture, on a 'first come, first served' basis. Each user tries to extract as much as possible from the resource in the fear that other users will exploit the resource first, and also in the belief that the amount they themselves use is only a small proportion of the overall stock. The consequences of ignoring the user costs are that the costs of extraction are undervalued, which results in exploitation rates that exceed the optimum. This is in contrast to the situation where a single user has rights to a resource: the user has to take user costs into account because it is the user who faces the increased costs of extracting from a depleted resource in the future.

Economic Valuation of Water Resources

Economic valuation here serves as a basis for evaluating the trade-offs involved in the allocation of water resources between competing wants. The focus is particularly on valuation of changes in functions provided by water resources under different allocation options. For the purposes of cost-benefit analysis, the impacts of alternative options for water resource use or management are specified in terms of the economic value using the common numeraire of money. Economic value is determined by the impact on social welfare, which is given by the aggregate impact on the utility of individuals in society.

The utility to individuals is determined by their preferences, which individuals express in the amount that they are willing to pay for goods and services. In addition to considering the economic value of water in terms of the common numeraire of money, the value of water also needs to be commensurable in terms of place, form and time. Water is a 'bulky' resource with high conveyance costs and, hence, its value may differ with location. Demand for water can also vary greatly over time. Thus, comparisons of value should ideally be in terms of raw water supplies at a specified point of diversion.

Linking Functions and Service Flows

Water resources are natural assets that create flows of goods and services over time. The key to valuation of water resources is to establish the functions that they provide, i.e. the link between the structures and processes of water resources and the goods and services they provide that are valued by society. If that link can be established, then the concept of derived demand can be applied. The value of a change in the functions provided by a water resource can be derived from the change in the value of the stream of goods and services provided.

The goods and services can be categorised in various ways, for example, in terms of whether they are extractive or *in situ*. They are influenced by extraction and return flows, which affect the quantity and quality of water stocks and flows. These influences relate in an intertemporal way to the stream of goods and services provided, and require incorporation into any meaningful valuation analysis.

Rogers, Bhatia and Huber consider the value of water to be divided into economic value and intrinsic value. Turner and Postle consider the economic value of water resources and aquatic ecosystems in terms of four separate components, and Young distinguishes between five categories of water-related economic values and also considers the possibility of certain other value types. De Groot categorises the components of ecosystem value according to the impact on welfare, using a broad definition that encompasses environmental, physical and mental health, employment and social contacts as well as material prosperity.

The approach advocated here is to describe the components of the value of water using the conventional categories of total economic value (TEV). There are two main categories, use values and non-use values:

— Direct use values arise from direct interaction with water resources. They may be consumptive, such as use of water for irrigation or the harvesting of fish, or they may be non-consumptive such as recreational swimming, or the aesthetic value of enjoying a view. It is also possible that 'distant use' value can be derived through the media, although the extent to which this is attributable to a specific site, and the extent to which it is actually a use value, are unclear.

— Indirect use values are associated with services provided by water resources but that do not entail direct interaction. For example, they are derived from flood protection provided by wetlands or the removal of pollutants by aquifer recharge.

Non-use values are derived from the knowledge that a resource is maintained. By definition, they are not associated with use of the resource or tangible benefits that can be derived from it. Non-use values are linked to ethical concerns and altruistic preferences, although it can be argued that these ultimately stem from self-interest. They can be divided into three types of value: existence value, bequest value and philanthropic value.

Existence value is the satisfaction derived from knowledge that a feature of a water resource continues to exist, regardless of whether or not it might be of benefit to others. Bequest value is derived from the knowledge that a feature of a water resource will be passed on to future generations so that they will have the opportunity to enjoy it. Philanthropic value is the satisfaction gained from ensuring that resources are available to contemporaries in the current generation.

There are two further types of value that are not categorised as either use or non-use values. These are option value and quasi-option value:

— Option value is the satisfaction that an individual derives from the ensuring that a resource is available for the future given that the future availability of the resource is uncertain. It can be regarded as insurance for possible future demand for the resource.

— Quasi-option value is derived from the potential benefits of waiting for improved information prior to giving up the option to preserve a resource for the future. This is based on a desire to take advantage of the prospect of improved information in the future and act on subsequent revision of preferences. It is the value placed on retaining flexibility, and on avoiding irreversible damage that might prove to be undesirable in the light of future information. An example is the value placed on conservation of a wetland until further information is available on the value of the species that are found within it.

Use of TEV in the analysis of alternative allocations ensures that the full social benefit of goods and services provided by water is taken into account. This is necessary to indicate to decision-makers the welfare improvement that is offered by alternative allocations. However, TEV does not provide an exhaustive assessment of the value of water resources to society. It measures the extent to which goods and services provided by water touch on the welfare of society, as direct determinants of individuals' wellbeing or via production processes.

It represents two fundamental sets of values: individual values and production values. Individual values include recreational and amenity values, as well as non-use values (existence, bequest and philanthropic values) of goods and services provided by water. Production values occur through the influence of water on the production and cost functions of other marketed goods and services (such as use of water as an intermediate good in irrigated crop production). The effects of this influence on the prices of other inputs and marketed goods and services translate into changes in individuals' welfare.

However, another set of values is supplementary to TEV. This represents the role of water resources in natural systems. It includes the value of services that stabilise natural systems and perform protective and supportive roles for economic systems. These values are more usually presented in relation to biodiversity, but are similarly applicable to water resources. They include the following (somewhat overlapping) categories of value:

— *inherent value*: the value of those services without which there would not be the goods and services provided by the system;

— *contributory value*: this represents the economic-ecological importance of species diversity. Species that are not of use to humans are important because they contribute to increased diversity, which itself contributes to the generation of more species;

— *indirect use value*: this is related to the support and protection provided to economic activity by regulatory environmental services;

— *primary value:* incorporates the fact that existence of the catchment structure is prior to the range of function/good and service values;

— *infrastructure value:* this relates to a minimum level of ecosystem 'infrastructure' as a contributor to its total value.

These values build on three important aspects of the ecology of natural systems:

— *Complementary relationships.* Species coexist within natural systems, defined by complex relationships of interaction and interdependence. Survival of one species depends on the existence of other species, which in turn depend on others. This 'contributory value' focuses on the survival of species within the web of interactive relationships with each species contributing to the survival of others. Contributory value is based on the limited substitutability of species. This occurs because every species performs very specific duties within the ecological system. The role of contributory values is not usually taken into account explicitly, because the required knowledge (on ecological interrelationships) is unavailable, but it can be incorporated through adoption of a precautionary approach to resource management.

— *Keystone species.* The persistence of natural systems in their current existing states may be dependent on a limited number of biotic and physical processes. These processes are directed by groups of species with complementary functions, known as 'keystone species'. Other species are redundant, although they can become keystone species under a change in environmental conditions. As long as species can substitute for each other under changing conditions, the balance of processes within the system can remain intact. However, reductions in the diversity of species in the system diminish the possibilities for substitutions under a change in conditions. This limits the capacity of the system to persist in its current state in the face of stresses and shocks.

— *Goods and services provided by a natural system are dependent on the structure and functioning of the systems.* The goods and services provided are connected inherently to the integrity of the natural system and the totality of the structure and functioning of the system. This can be understood in terms of the concepts of primary and secondary value. The primary value describes the system characteristics: the self-organising capacity of the system including its dynamic evolutionary processes and capacity to absorb external disturbances. It relates to the aspects of the system that "hold everything together" and is consequently also referred to as "glue value". Secondary value refers to the renewable flow of goods and services generated by the natural system. It is dependent on the continued operation, maintenance and 'health' of the system as a whole.

TEV does not give credit to this set of values and, therefore, is not exhaustive. Such values are particularly relevant to single function natural systems, the contributory value of which can only be addressed properly when the site is viewed within the context of the larger catchment system.

The recognition of complementary relationships implies that the total value of catchment systems is infinite. This is similar to the consideration of water resources as a form of 'critical' natural capital. Here again, the value of water is infinite and the usual measures of value do not reflect the true economic value of the resource. As a basis of human life, complementary relationships with water resources are indispensable under realistic technological and economic conditions.

However, apparently marginal decisions are important in the real world and, therefore, need to be considered. The problem is that knowledge about the consequences of resultant infringements on natural systems is incomplete. There is an unbridgeable gap in knowledge about natural system interrelationships and regularities. The benefits of protection will often only become apparent once the natural system has been disturbed or lost.

The task of sustainable management can be defined as sustainable utilisation of the multiple goods and services generated by natural systems, together with "socially equitable" distribution of welfare gains and losses inherent in such usage. However, social welfare is affected both by changes in economic welfare and also changes in properties of natural resources that are associated with people's sense of identity, their culture and which are of historical significance. Such properties are particularly important in the case of water resources, given the essential role of water for human life. The compilation of data for such properties is a qualitative exercise, involving more deliberative and inclusionary interest group approaches, such as consensus conferences, citizen juries and focus group interviewing.

Different cultural views on social relations are assumed to give rise to different degrees of support for alternative decision-making procedures and for the underlying valuations elicited via the social discourse process. This has similarities to the so-called 'approved process' approach in which all relevant parties observe a specified set of procedures or concept of due process to make a decision that balances conflicting values at the political level.

Some environmental analysts claim that natural systems also have non-anthropocentric intrinsic value and that non-human species possess moral interests or rights, or that although all values are anthropocentric and usually instrumental, the economic approach to valuation is only partial. These environmentalist positions lead to the advocacy of environmental sustainability standards or constraints, which to some extent obviate the need for valuation of specific components of the environment. However, it is still necessary to quantify the opportunity costs of such standards, or to quantify the costs of current, and prospective environmental protection and maintenance measures. Nevertheless, some commentators view it as feasible and desirable to manage the environment without prices. For example, O'Neil found that in other arenas such as forestry and biodiversity management, issues concerning conflicts in value are resolved through pragmatic methods of argument between botanists, ornithologists, zoologists, landscape managers, members of the local community, and farmers.

A growing body of evidence suggests that some of the conventional economic axioms are violated systematically by humans in controlled experiments and in everyday life. To take just one issue, it seems likely that individuals recognise 'social interest' and hold social preferences separate from self-interested private preferences. The origin of this social interest may be explained by theories of reciprocal altruism, mutual coercion, or by sociobiological factors. Therefore, the distinction between the individual as a citizen and as a consumer is not an 'either/or' issue, but is more properly interpreted as the adoption of multidimensional roles by individuals.

As citizens, individuals are influenced by held values, attitudes, and beliefs about public goods and their provision. In this context, property rights (actual and perceived), social choices and moral concerns can all be involved in the conflict between conservation and development of natural resources. The polar view to the conventional economic approach holds that the very treatment of ecological assets such as biodiversity in terms of commercial norms is itself part of the environmental crisis.

The argument becomes one of the 'proper' extent of market influences and commodification. Advocates of this perspective argue that market boundaries should not cover as many environmental assets as possible. Instead, society should give greater consideration to the nature of deliberative institutions for resolving environmental problems and the social and economic framework that sustains them.

A counterbalancing argument is that some environmental goods and services that have mixed public and private good characteristics (e.g. forests, catchments, areas with ecotourism potential and some aspects of biodiversity services) could be privatised or securitised (shares issued). In this way, self-interest and the profit motive can be made to work in favour of environmental conservation.

Use of Economic Values Water Resources Management

The quantification of the economic value of water resources and the identification of those instrumental values that it is not possible to quantify is of importance to the management of water resources for the five reasons presented below.

The importance of water in national development strategies. The depletion and degradation of water resources imposes costs on nations, some of which affect the gross national product (GNP). Typically, the degradation of water resources contributes towards a reduction in GNP, whereas resource depletion contributes towards increases in GNP. However, GNP accounts do not include the costs imposed on society by resource depletion and degradation; although they would be included if GNP reflected more comprehensive measures of aggregate well-being. Although the empirical investigation of water resource depletion and degradation is in its infancy, the evidence available suggests that the costs of resource depletion and degradation are appreciable. Such estimates of resource costs estimates can play a useful role in assessing development priorities. As the costs of resource depletion and degradation are increasingly recorded and accorded greater significance, planners have greater incentives to prioritise these issues in their development plans.

Modification of national accounts. national accounts arc deficient in the treatment given to environmental resources, such as water. Measures of economic activity ignore the resource flows that take place in the economy. These fail to record important activities that affect the sustainability of the economy and of well-being. Thus, there is a need to modify national accounts such that they record "stocks" and "flows" of natural resources. GNP should account for depreciation of resource stocks in the same manner that it incorporates depreciation of human-induced capital. This would provide a measure of the 'draw down' on water 'capital' and the losses that accrue to human well-being from the use of goods and services provided by water resources. Both adjustments involve economic valuation, although national accountants have not agreed how best to make the appropriate adjustments.

The setting of national and sectoral priorities. Information on the economic value of changes in water policy can assist governments in setting policy and sectoral priorities. A comparison of the benefits and of the costs of planned changes in policy is required in order to establish whether they are potentially worthwhile. Valuation can be used to influence the allocation of irrigation water, ensuring that water is directed towards priority areas. For example, decision-makers can use it as an aid in the allocation of water between hydroelectric power generation and storage for irrigated agriculture. There is a particular need to review sectoral priorities in terms of economic benefits and costs, which has perhaps even greater force in the developing countries where government income is at a premium.

Project, programme and policy evaluation. Environmental resource damage and benefit estimation falls conventionally under the remit of project appraisal. Extension of project appraisal to account for impacts of water resource degradation and depletion presents no conceptual problem for the benefit-cost approaches that are used. It is important that the environmental implications of projects and programmes be evaluated. Indeed, the overall returns to development programmes should be assessed with reference to environmental enhancement components.

Investments in water resources are important components of public infrastructure budgets, and include irrigation, hydropower, urban and rural water supply, sanitation and flood control. Valuation is also employed in the assessment and implementation of policies that are used to monitor and manage water resource depletion and degradation. The policies can include standards that require the polluter to bear costs in meeting the standard that are equal to the minimum estimated value of the damage that the pollution would cause. Valuation is also an important guide in the setting of environmental 'prices' in the form of taxes, charges or tradable permits. However, the accuracy and reliability requirements of monetary valuation results are much stricter for this purpose. Large estimation errors are usually unacceptable in view of the political sensitivity and potential consequences if tariff setting does not achieve the intended effect because it is based on the wrong information.

Economic valuation and sustainable development. Economically efficient use of water resources is not necessarily sustainable. For example, the optimal rate at which a finite non-renewable resource is depleted requires a positive rate of extraction. In the absence of discovery of further identical resources, the resource must be exhausted eventually. Every unit of resource used today is at the cost of foregone use of a unit tomorrow. This is relevant to overextracted reserves of groundwater and poses the question of how much groundwater to pump now and how much to save for future needs. On the other hand, sustainability can be interpreted as a requirement that human well-being does not decline through time. Therefore, adoption of sustainable development as a goal creates a need for economic valuation to establish that human well-being does not decline through time.

With 'weak' interpretation of the concept, sustainability can be defined such that the primary condition is for the aggregate stock of capital not to decline. This requires valuation of the capital stock to establish the extent of the stock and to monitor whether it is in decline. Thus, valuation can support pursuit of sustainability at the very least by helping to focus policy-maker and public attention on threatened resources. However, the economic valuation of water resources is a crude and inexact science. The value of water varies widely according to factors such as the use it is put to, the socio-economic characteristics of users, its availability in space and time, as well as the quality and reliability of supply. It is not proposed that technocratic decisions on allocation should be made solely on their basis of estimates, or that they should be made in a routine fashion. Rather, it is proposed that the estimates obtained are useful for highlighting more general themes in water use that have major implications for policy.

Bibliography

Baumann, D. D. and Dworkin, D. M. (eds), *Planning for water reuse*, Maaroufa Press, Chicago, 1978.

Bouwer, H., "Integrated water management: Emerging issues and challenges", *Journal of Agricultural Water Management*, 2000.

Brown, L. R., "How water scarcity will shape the new century", *Water Science and Technology*, 2001.

Charney, A. H. and Woodward, G. C, "Socioeconomic impacts of water farming on rural areas of origin in Arizona", *American Journal of Agricultural Economics*, 1990.

Frederick, K. D., "America's water supply: Status and prospects for the future", *Consequences*, 1995.

Gale, J., Line, D., Osmond, D., Coffey, S., Spooner, J., Arnold, J., Hoban, T. and Wimberley, R., "Evaluation of the experimental Rural Clean Water Program", United States Environmental Protection Agency, 1993.

Gelt, J., Henderson, J., Seasholes, K., Tellman, B. and Woodard, G., "Water in the Tucson area: Seeking sustainability", 1998.

Gleik, P. H., *The World's Water 2000-2001*, Island Press, Washington, D.C., 2000.

Gliessman, S.R., "What are the indicators of sustainability in agricultural systems?" In *Proceedings, "Organic '92" Farming Symposium. University of California*, 1996.

Goldberg, R., "Experts try to predict the future cost of water", *Proceedings of the Fifth National Conference on Environmental Issues, Water: Our Next Crisis*, Academy of Natural Sciences, 1994.

Gollehon, N. R., "Water markets: Implications for rural areas of the West", *Rural Development Perspectives*, 1999.

Hetzel, G., "Safe use of pesticides in agriculture, Virginia Cooperative Extension, 1996,

Hillel, D., *The Efficient Use of Water in Irrigation*, World Bank, Washington, D.C., 1987.

Los Huertos, M., L. E. Gentry, and C. Shennan, "Land use and instream nitrogen concentration in coastal agricultural watersheds". *In Optimizing Nitrogen Management in Food and Energy Production and Environmental Protection: Proceedings of the 2nd International Nitrogen Conference on Science and Policy*, 2001.

Mitchell, J. P., C. Shennan., M. J. Singer, D. Peter, R .O. Miller, T. Prichard, S. R. Grattan, J. D. Rhoades, D. M. May, and D. S. Munk. "Impacts of gypsum and winter cover crops on soil physical properties and crop productivity when irrigated with saline water", *Agricultural Water Management*,2000.

_______________ ., D.W. Peters, and C. Shennan, "Changes in soil water storage in winter fallowed and cover cropped soils", *Journal of Sustainable Agriculture*, 1999.

Werner, Mattnew R., "Water, water everywhere: Managing organic matter is key to improving soggy soil", *Farmer to Farmer*, 1995.

Index